Historic Mills of America

Landmark Publishing, Inc.
4410 W. Vickery Blvd., Suite 101
Fort Worth, Texas 76107

Library of Congress Data:
Fralish, Lynda, Editor
Historic Mills of America
Landmark Publishing, Inc.

Includes Index
ISBN 0-9667999-2

First Edition

Landmark Publishing, Inc. books are available at special discounts for bulk purchase for sales promotions, premiums, fund-raising, or educational use. For details contact:

LANDMARK PUBLISHING, INC.

PRESERVING HISTORY

4410 West Vickery, Suite 101 - Fort Worth, Texas 76107
1-800-886-0689
1-817-377-3399

Printed in the United States of America

The Society for the Preservation of Old Mills

The Society for the Preservation of Old Mills, also known as SPOOM, was chartered in Maine as a non-profit organization in 1972. Its members include mill owners, old mill buffs, museum curators, conservators, writers, teachers, artists, photographers, equipment supply firms, and institutions such as libraries. Membership is open to any individual who shares the interests and purposes of SPOOM.

The Society is an international organization, which is dedicated by its constitution to:

> "promote interest in old mills and other Americana now quickly passing from the present scene. It reports to its members through a quarterly magazine... keeps files and maintains a library on mills...and acts as a clearing house on milling information among all those interested."

The organization publishes a magazine, "Old Mill News," on a quarterly basis. It describes various mill restoration projects, along with technical information and organizational activities. It has become an excellent source of mill information. In addition we also reprint older out of print books, and have them for sale in our Bookstore.

Currently there are three local chapters, located in New England, Canada and The Great Lakes. Each chapter holds two meetings a year at various mills in their area. These meetings are a great place to meet new friends with milling interests and visit mills that may not be open to the public on a regular basis.

We also hold an annual three-day meeting throughout the country. Generally, there are seminars on Friday, followed by a bus trip to visit various local mills on Saturday. A "Meet the Board" event is held on Sunday morning and other mills in the area are open to people who want to stay and visit them.

Information about SPOOM membership can be found on our website
www.spoom.org
or obtained by writing to
William L. Denton
5444 Alpine Ridge
Stevensville, MI 49127-1302.

Photo by J.C. Barger

Acknowledgements

Mills? You mean those buildings with the big waterwheels? I saw a photo of one once in a book! Are there any still around?

Little did I know that the creation of this book would take me on a journey to hidden places unknown to many. With the help of the Society for the Preservation of Old Mills, a wonderful opportunity to walk in the paths of history and beauty is offered within these pages. The experience has been a unique education in my life that will never be forgotten.

Stories and beautiful photographs came from mill owners all over the United States and each had an amazing story to tell. Owners of sawmills, flour mills, gristmills, woolen mills, textile mills, plaster mills, cotton mills, mustard mills, tortilla mills and mills to provide electricity responded. Everything known about each mill was shared, from history to mill functions, and even a few ghost stories.

As you wander through the mills in the following pages, be sure to take time to imagine taking your crop to the mill and waiting "your turn". Smell the aroma of the different grains, feel the texture of the fabrics made there, and try the "rule of thumb" on the flour to see if you know when to adjust the millstones.

If you listen carefully you may hear the singing of a waterwheel from the past, the rushing of the water over the flume, or the voices of the farmers visiting at the mill. The laughter of children playing in the stream and families listening to a band while sharing a picnic on homemade quilts in the grass may also be heard. Can you hear the war cry of the Indians or the gunfire of the Civil War?

Look around you and envision the millwrights, millers, slaves and families who struggled and labored to keep the communities fed. Ride by the coffin maker's mill and look in the window to see if a reed there tells of someone who has died and check the length of it to determine if it is an adult or child. In some of the mills the spirits of a judge sentencing an outlaw, or a nurse bathing a wounded or ill person may still be around. Become familiar with the men and women of the past with their hardships and varied personalities. You will find rogues, envisionaries, heroes and star-crossed lovers.

It would be virtually impossible to thank each and every person who contributed to this book but I would like to acknowledge a few who helped thoughout its creation.

Thanks to many photographers who allowed us to publish the unique photos pictured here. My thanks to J.C. Barger, Palmer Werner, and Steven Spring for allowing us to use photographs from their extensive collections. A special thanks to Joe Seme, who not only allowed us to use his photographs but also furnished us with two of his beautiful watercolors.

Thanks, also to the wonderful individuals and groups that fought to keep history alive for us. They have been gracious enough to entrust us with their stories and photographs, without which there would have been no book. Please take the time to read each name and consider the effort it took to restore and save these priceless treasures.

And of course, a very special thanks to Sidney Halma and his associates at the Society for the Preservation of Old Mills, for all of their support.

It would be negligent not to thank my fellow staff members, including Ernie Hammond, Operations Manager; Paula Oates, Senior Editor; Jason Petty, Director of Marketing; Lisa Blake, Administrative Assistant; and Julie Guess, Production Assistant. Thanks also to our President, Robert Akin, Jr., for allowing us to take time to do the many hours of research needed to finalize production of *Historic Mills of America*.

Lynda Fralish, Editor

HISTORIC MILLS
of AMERICA

Aderholdt Mill
Alabama

In the spring of 1832 Col. Dick Williams, using slave labor, began construction of the Aderholdt Mill on the Little Tallassechatchee Creek. Completed in 1835, power for the mill came from a quarry opened near the creek. Huge boulders of blue granite were constructed into a stone dam resulting in a mill pond. Kilns burned day and night baking the brick, hand formed from the red Alabama clay.

The mill had brick walls more than a foot thick and an overshot waterwheel measuring twenty-three feet in diameter, almost the height of the building. An unbroken length of fifty feet from one end of the building to the other are exposed hewn beams. The mill rocks were lifted into place and the remaining machinery installed. Construction costs for mill and dam were approximately $2,000.

The mill was now ready to grind corn and wheat, crush feed, thrash peas and beans, saw lumber, make syrup, gin cotton, and build furniture and caskets for the local community.

Undiscovered by General Sherman on his "Fiery March to the Sea", the mill was spared destruction during the Civil War.

Col. James Ainsly Stevenson acquired the mill in the early 1840s. In 1872 it passed from Stevenson to I. Seymour and his wife, Olivia, and a short time later to Alex O. Stewart. A few months later Stewart sold it to General William H. Forney. Forney retained the mill until his death after the Civil War.

Daniel Hoke Aderholdt was one of the millers hired to work the mill during this time. One day his new bride, while bringing his lunch to the mill, stepped too close to a shaft. Her sweeping skirts were caught, pulling her to her death.

Other deaths occurred at the mill, a young girl lost her footing on the slippery millrace and fell to her death on the rocks below, and one angry boy pushed his younger brother from a third floor window.

In 1864, after Forney's death, the mill and sixty-five acres sold at auction in front of the Jacksonville courthouse. Money was scarce and the mill sold for $1,100 to Francis M. Treadaway. Robert Treadaway inherited the mill in 1902. In 1912 James Emanuel Aderholdt and his son, Jim, purchased the mill. It then became known as the Aderholdt Mill.

James Emanuel Aderholdt came from a heritage of millers. His father, Abraham, moved his family and their belongings from South Carolina to Alabama in a covered wagon in 1838. James barely escaped starvation during the Civil War and, thereafter, never tired of grinding grain to feed others. His son, Jim, ran the lathes, sawmills, sorghum mills and cabinet shop. When Emanuel died, Jim's sons helped run the mill. Jim known as "Jim Daddy", replaced the lumbering one hundred-year-old waterwheel with a modern turbine wheel in 1934. "Jim Daddy" continued to run the mill until the age of ninety-four, in 1976, grinding meal up until the last six weeks of his life.

In 1995, Bud and Alice Sawatsky fell in love with the old Aderholdt Mill. They purchased it and spent two years on restoration. The first floor is now living quarters but the second and third levels remain as they did when the mill was fully operational. The Aderholdt Mill is listed on the National Register of Historic Places, 1836. Restoration work continues and an eighteen foot overshot wheel has been installed. The Sawatskys hope to re-open their wonderful old "Aderholdt Mill".

Story and photo courtesy Bud and Alice Sawatsky.

Beans Mill

Alabama

Indications are that mills of one kind or another have existed on this site since 1836.

During the Civil War a gristmill and an up-down sawmill were in operation on opposite sides of Halawaka Creek, sharing the same dam. Court documents confirm the gristmill was "wrecked" in an 1874 flood. Apparently the sawmill and dam were spared as they were advertised for sale, along with "a lot of damaged machinery from the wreck of the mill." It seems important lessons were learned from this disaster. A new mill was built, not in the creek as the earlier one, but alongside. Stone piers were placed on bedrock, tall enough to support the building above high water.

The new mill was apparently powered by a pre-Civil War turbine salvaged from the original mill. Remnants of this old turbine remain. The mill was set up to grind both wheat and corn and would have been considered a merchant mill.

When George Bean bought the mill from Ben Franklin Meadows in 1903, the turbine was apparently on its last legs. Sometime before 1910, a fourteen-foot diameter, six-foot wide overshot waterwheel was purchased from Fitz Waterwheel Co. and installed at what had become Beans Mill.

One time a wall of water from a failed dam upstream smashed the dam, but the operation survived until 1952 when Mr. Bean died.

The abandoned mill deteriorated for almost forty years, until purchased by John and Faye Ross, along with eighty acres, in 1989. By this time the structure had reached a point where it was not realistic to save it except by reconstruction. The building was dismantled and salvageable timbers identified for reuse.

In 1995 after almost five years of storage, the timbers were brought together as a structure once more. A waterwheel was found in Georgia and renewed the reconstruction efforts. Heart pine timbers from an old textile mill were donated, also time and equipment of a sawyer who wanted to help. Several windows were donated, and much time and support came from volunteers. On December 4, 1997, corn was ground for the first time in forty-five years at Beans Mill.

One of the most pleasant surprises in the restoration of Beans Mill occurred after the owner's return from a SPOOM conference in Cornwall, Ontario, Canada in September, 1997. Attached to the Ross' door was a business card from Frank Smith, a welding instructor for the local Junior College, with a note requesting the owners contact him. Through another instructor Mr. Smith had learned the Ross' were about to rebuild a waterwheel which would require considerable welding and he thought it might make a good project for some of his students. When Frank offered the facilities of the college and his students to weld the wheel with state of the art equipment and techniques, the owners did not hesitate to give the go ahead. The students were pleased and anxious to be working on something which would have value and long life in a historical restoration. One even bragged about one day being able to "tell his grandchildren about making that old waterwheel."

The mill is on Alabama's list of Historic Places. As of June 1999, the mill has become sort of a mini-museum and work is under way to install an antique turbine-generator to make electricity. When Mr. Ross is asked, "When will you be finished with the mill?", he answers, "probably never but come see us anyway." For more information call (334) 705-0773.

Story and photo courtesy John Ross.

Mountain Brook Mill
Alabama

Birmingham (Jefferson County), AL
Photo by J.C. Barger

Rikard's Mill

Alabama

Rikard's Mill is a restored, operational, waterpowered gristmill nestled in the piney woods along the banks of Flat Creek, which flows through Wilcox and Monroe Counties. With the gristmill as the centerpiece, the historical, eight-acre park serves as a museum dedicated to the preservation of past folk traditions, such as milling, blacksmithing, cane syrup making, and more.

First built in 1845, Rikard's Mill offered a center of activity. People from miles around would gather at the mill, bringing a "turn" of corn for grinding. While there, locals caught up on the latest news, had picnics and waded in the stream.

Jacob "Jake" Rikard (1806-1880), the builder, first operated the gristmill. A blacksmith, locksmith and mechanical genius, it is said he could "fix anything". He decided to design a water-powered gristmill to provide a good future for his large family.

In the early 1840s, in building the first mill, Rikard used logs for the millhouse construction. The power to drive the forty-eight inch millstones came from a wooden wheel, commonly known as a "tub" wheel, lying horizontally between two timbers directly below the mill. By 1845, Flat Creek washed the mill from its wooden pilings, whisking away Rikard's livelihood.

The year 1858 found Jake racing to complete the present mill before the War Between the States. He and two slaves, Stokes and Edward, built the present mill using board-and-batten construction for the millhouse, and native cypress to complete the roof. The huge forty-eight inch millstones were installed to run with the tub wheel.

After Rikard's death, the mill passed to his daughter, Mary and her husband, Holly Middleton Forte, then to their son, John Dennis Forte. In 1905, he installed a new metal Leffel turbine. He became full-time millwright after retiring from teaching in 1921. By the 1930s, the old millstones were replaced with the thirty-six inch stones used today. Prior to his death in 1966, John passed the mill to his son, Maurice.

In 1993, descendants of Jake Rikard, Mack Albert Rikard and Maurice Forte, agreed that the legacy of their family and Monroe County must be preserved. Mack Albert took on the restoration of the mill and dam. Thanks to his efforts, corn is ground into cornmeal and grits as in the past.

The people of Monroe County graciously donated materials and labor to complete other buildings in the park, which include a covered bridge, gift shop, working blacksmith shop, barn with nineteenth and twentieth century equipment, a mule-driven nineteenth century cane syrup mill and a carriage house. Construction of a cabin began in 1999 to provide an "outdoor schoolroom" for children's workshops and adult seminars.

Rikard's Mill is a Monroe County Property listed on the Alabama Register of Historical Sites. For more information, contact Rikard's Mill at (334) 789-2781 or Monroe County Heritage Museums at (334) 575-7433.

Story and color photo by Sandra K. Standridge. Historical photo by Aaron White.

Johnson Mill
Arkansas

This plot of land has always been associated with the Johnson Mill. At the time of the original survey in 1834, the Truesdale family occupied the site. Survey notes indicate the existence of the "Truesdale mill about three chains to the left" and "an old mill to the right about 200 links". If land surveyors in 1834 considered the mill "old" at this time, the approximate time the original mill was built is estimated to have been prior to the turn of the century.

After the filing of the first claim, in 1838, ownership changed many times. The Arkansas Products of Industry census for 1860 lists the mill as Sutton's Mill with a value of $10,000 and production of over $40,000 worth of grain annually: astonishing figures for this time period. Flood plains such as this site were valued for the high nutrient content and exceptionally deep soil.

During the Civil War Confederate soldiers retreating from the battle of Pea Ridge were again met by Union soldiers and a battle took place on the flood plain adjacent to the mill. When they retreated, the Confederate soldiers burned down Sutton Mill to prevent Union troops from capturing and using the grain stored there. The site remained unoccupied until after the war. At that time construction of a new mill took place.

In 1865, a deed was filed transferring the mill property to Jacob Q. Johnson. The new mill took two years to build with completion in 1867. Johnson bought out investors, William, Samuel and Lucinda Mayes in 1880, becoming the sole proprietor of the land and mill.

In 1977, M.F. Johnson fractured his hip and was forced to close the mill. The mill is considered one of the oldest such mills in the United States to be kept in continuous production under the same family ownership and management (approximately 110 years). Deeded perpetual water rights for the mill, from a spring originating northeast of the property, states owners of the spring cannot harm the water or interrupt its flow.

Johnson Mill underwent restoration completed in 1985. Another renovation took place in 1991. The mill is now a Clarion Carriage House Inn. Johnson Mill is listed on the National Register of Historic Places. For more information call (501) 443-1800.

Story and photo courtesy James Lambeth.

THE OLD MILL, PUGH'S MILL
ARKANSAS

15—"Interior View of Old Mill," Lakewood, Little Rock, Arkansas

5A-H517

5—The Old Mill, Lakewood, Little Rock, Ark.

8A-H2367

In 1933, developer Justin Matthews of the Lakewood area in Park Hill, contracted for the construction of a replica of an old waterpowered gristmill. Neither the stone mill building nor the general layout is a reproduction of any particular mill, but a familiar design of early Arkansas, designed to fit the contour of a rough area. The mill is intended to appear neglected - absent of doors and windows due to thieves or decay - just as old mills in service in the early 1800s had become by the 1930s.

Filmed in the opening scenes of the 1939 classic movie "Gone With The Wind," the Old Mill, honored on the fiftieth anniversary of the movie, also became the site of the unveiling of the "Gone With The Wind" commemorative stamp.

The park features rustic concrete sculptures in the shapes of footbridges, benches and trees designed by the nationally-acclaimed sculptor and artist Senor Dionico Rodriguez.

Several items in the Old Mill are actual historic relics. There is a gristmill on the first floor that dates back to 1828. The mill-rocks, now used for seats, have "bear picking" marks (deepening of furrows in the face to increase their grinding capacity), and date back to 1840 and 1923. Also found are mill rocks on the second floor from the plantation of Tom Knoble, the grandfather of Tom Pugh, in whose honor the mill is dedicated, and two original mile stones from a road laid out more than 150 years ago by Jefferson Davis.

Along this road, the Cherokee and Choctaw Indians traveled from the present town of Dardanelle into the Indian Territory, now Oklahoma. Three sections of a wrought-iron shaft cut from the stern wheel of a passenger steamboat that traveled the Arkansas River in the 1800s were used for construction of the "broken tree" bench.

The Old Mill, a gift to the City of North Little Rock by the former Justin Matthews Company in 1976, was added in 1986 to the National Register of Historic Places.

The Old Mill is now part of a city park maintained and landscaped by the Pulaski County Master Gardeners and North Little Rock Parks and Recreation. For more information call the North Little Rock Visitors Center at (501) 758-1424 or (800) 643-4690.

Photo and story courtesy
North Little Rock Advertising & Promotion Commission.

Post cards courtesy of the
Society for the Preservation of Old Mills.

War Eagle Mill

Arkansas

Located on the War Eagle River, east of Rogers, Arkansas, is the War Eagle Mill.

Sylvanus and Catherine Blackburn built the first mill and a dam constructed of logs and stones in the 1830s. After moving from Tennessee to Arkansas, they cleared land, planted crops, built a two-story house, carpentry shop, soap house, smokehouse, blacksmith shop and all workings necessary to a pioneer settlement. Catherine bore eight children and adopted eight.

Farmers came from long distances to have grain ground. The mill stayed busy and proved successful. The first four-story mill washed away in 1848.

Sylvanus built a second mill, but it was burned during the Civil War in 1862 by the order of a Confederate general, to prevent capture and use by the Union Army. Both sides had ground corn at the mill for the troops.

James A.C. Blackburn, son of Sylvanus, rebuilt the War Eagle Mill in 1873. By now, War Eagle had become a thriving settlement with two general stores, a school and a Masonic lodge. Two steel roller mills ran flour and a set of stone buhrs ground corn. In 1924 the cry of "fire" rang throughout the village and the gristmill mysteriously burned.

Jewell and Leta Medlin and daughter, Zoe Medlin, built the current working waterpowered gristmill, in 1973, on the original foundation. Zoe Leta Caywood, as milleress, is owner and operator.

An authentic reproduction of the 1873 mill, the new gristmill grinds corn, wheat, rye and buckwheat. An eighteen foot undershot waterwheel powers a thirty inch run of stone buhrs. The waterwheel is made of Arkansas cypress wood. The first new gristmill built in Arkansas in one hundred years, it has the only working undershot waterwheel in the United States. The second floor of the mill is a gift shop. The third floor houses a restaurant.

In 1924, when the last War Eagle Mill burned, there were 124 registered gristmills in the State of Arkansas. Today there is only one. For more information call (501) 789-5343.

Story and photo courtesy Zoe Caywood.

Bale Gristmill

California

When settlers began to arrive in California's Napa Valley in the 1830s and 1840s, they planted crops such as corn and wheat to replace the oats they found growing wild. To convert these grains into usable flour they required a mill, which became a social center as people gathered to visit, "mill around", while their grain was being ground. The miller, respected for his skill, was a leading citizen and his counsel on banking and business matters sought and frequently paid for.

Construction on the gristmill and granary began in 1846. Local materials were used, including Douglas Fir and Coast Redwood cut from nearby canyons and hills. Some timbers were cut to length with the bark left on, while others were roughed out with hand tools. Some lumber may have been obtained from the waterpowered sawmill that the owner, Dr. Bale, built on the Napa River around the same time. Initially, two sets of stones made from local rock were used.

The timbers were generally notched and held in place with wooden pegs, in addition to some square nails and screws. The foundation is of the abundant native stones, laid dry. Indians provided much of the labor.

Dr. Bale, an English surgeon who arrived in California in 1836, was made surgeon for the Mexican army then in control of the territory. He was granted the northern half of Napa Valley as a Mexican Land Grant after he married Mexican General Vallejo's niece. He traded six hundred acres of this grant for metal work on the mill.

In 1847, quartzite buhrs that occurred in small deposits near Paris, France replaced the original stones. Today's miller continues to use the 1847 stones, which have a hardness equal to an emerald, and a life expectancy of 125 years of full time use.

Originally, an overshot waterwheel powered the mill's complete operation. Water came from Mill Creek, which runs directly south of the granary, into a millpond formed by damming an intermittent creek behind the mill.

The first wheel, twenty feet in diameter, did not provide enough power during dry summers and was replaced a few years later by a larger one, like the one seen today. The new wheel held more water, which fell farther, and generally produced enough power from the small water flow. The wheel, thirty-six and one half feet in diameter, weighed approximately seven and one half tons. During operation it made about three revolutions a minute.

A flume of cut lumber replaced the original flume, made of redwood logs split in half, hollowed out and laid end to end on trestles. In the 1860s a steam engine supplemented the waterwheel power, and in 1879 a still more efficient water turbine fed by a large metal pipe ended the operation of the wheel. The mill ceased operation in 1905.

In 1974 ownership passed to the California State Parks System and restoration to the 1860 configuration began. In October 1988 operation resumed with the rumble of the waterwheel heard once again. For further information call (707) 942-4575.

Story and Photo by Bob Bennett.

Postcard courtesy Society for Preservation of Old Mills

Santa Inés Mission Mills

California

As he penned the end-of-the-year report for 1817, Padre Francisco Xavier de la Conceptión Uría contemplated the state of Mission Santa Inés. On the one hand, the new church replacing the original structure shattered in the earthquake of 1812 had been dedicated in July. The harvest had been good with over 8,000 bushels of corn and wheat, in addition to beans and vegetables. Over 12,000 animals, mostly cattle and sheep, grazed peacefully on the meadows beneath the bluff and a neophyte population of 920, representing the largest number yet, lived at the Mission. On the other hand, the Mission was only restored to what it had been nine years before, when the Vizcayan priest, eager to work with the ingenious and adept Chumash Indians, had arrived. The earthquake had been a severe setback and in the intervening years the world had undergone drastic changes. The supply ships from Mexico, essential to the survival of the California frontier, ceased to sail in 1810. The missions were forced to support the military with supplies and funds, slowing the work with the neophytes and diverting resources that might have been better used. Further, secularization now seemed an ominous possibility.

Thus it was essential to push ahead and increase production in order to support the troops and bring technology and civilization to his Indian charges. While continuing to rebuild and extend the main Mission complex, Uría, in 1819, selected a site about a half mile from the church along the banks of Alamo Pintado Creek, for construction of a gristmill.

A massive work, it consisted of two connected reservoirs built into the slope of a hill with a mill structure at the lower end. The water was supplied by a small zanja (conduit) snaking along the edge of the first bench of the river for more than two miles. The mill was an ancient design, a horizontal wheel powered by a water jet that turned a stone attached to its axle. Crude, but appropriate to the water supply and the amount of grain to be ground. A fulling mill to take advantage of the endless supply of wool, something the padres had been talking about since the 1790s, required a miracle.

Uría's prayers were answered in the person of the American, Joseph Chapman. Captured in Monterey in late 1818 during the raid of the Argentine privateer, Hippolyte de Bouchard, Chapman earned parole when Governor Sola discovered his mechanical prowess. To Uría's delight, Chapman arrived at Santa Inés in December 1820, and within minutes, sketched out an ingenious plan for a fulling mill. Chapman supervised the construction of a typical new England overshot wheel, the power from which lifted a set of hammers that pounded and turned woolen cloth at the same time it was washed by a steady stream of fresh water. Uría now felt prepared to show the genius of European civilization to the natives and improve the economic status of the missions. That he had much of a chance to do so is doubtful.

In 1823-1824, the Chumash revolt stopped progress and badly damaged the Mission buildings. Uría was immediately transferred to Santa Barbara and, later, Soledad. Work with the natives slowed. Inevitably secularization overtook Mission Santa Inés, and the property divided in 1836, under orders of Governor Mariano Chico. Apparently the mill property and mission were separated, for there is no mention of it when the Mission property was returned to the Franciscan order. Thus the mill structures have stood for more than a century and a half, and through a combination of good luck and respectful owners, the entire complex has come to us intact, except for the ravages of time. The previous owners, Ellen and Harry Knill, comprehending the immense value to history and understanding what it represents, not only took care to preserve the site, but sponsored archaeological studies and roofed the fulling mill before transferring title to the Santa Barbara Trust for Historic Preservation.

The Santa Inés Mill Complex, not currently open, awaits a new destiny. For more information call (804) 965-0093.

Story by Richard Oglesby.
Story and photos courtesy of David Debs.

Westfair Mill

California

Westfair Mill was recreated from Boddington Mill, a sixteenth century mill built in Cotswold, England. The five-story edifice is constructed from the old timbers, plaster and brick of the original mill. This mill stands atop a six-acre promontory in Malibu, California, a tribute to several European legacies.

The entrance galley was once in Eton hall; the bar front, the London Palladium ticket booth; the music room door, a beveled glass made for Sarah Bernhardt. Victorian gold molds from Thomas Edison's European estate depict Electra; a large opalescent dome of Tiffany Favrile glass immortalizes a ceiling; a masterpiece by Rene Lalique that survived the war and became Coco Chanel's hideaway at the Hotel Ritz is the master bedroom; miller Thomas Crapper's sleeping chamber and bathtub have been left as they were found. Also on the grounds are a sixteenth century windmill and a mosscovered tower with an authentic crossbow slit.

The ground floor of Westfair Mill is a private museum housing the old stone, which still grinds corn to flour.

Courtesy Society for the Preservation of Old Mills.

Comer Water Wheel
Colorado

Coloradan John Comer didn't reinvent the wheel, but the rancher did restore the seventy-three-year-old wooden waterwheel on his spread near Vail. Powered by nothing more than the rushing Colorado River and gravity, the forty-six foot tall, ten foot wide wheel carries river water to a trough and pipe system, irrigating sixty acres of semiarid pastureland.

Earl Brook, the original landowner, devised this apparatus in 1922 to water his upland property. Today the wheel supplements an electric irrigation system. Comer, who secured listing for the wheel in the National Register of Historic Places in 1976, duplicated the original materials-redwood for water baskets and Oregon fir for spokes. Charles Keir, Brooks's eighty-two-year-old son-in-law, lent his expertise gained by helping Brooks refurbish the wheel in 1937. The Colorado Historic Preservation also pitched in with a $35,000 grant.

Designation in The National Register of Historic Places was awarded in 1976. Preservation News.

Photo and story courtesy of John Comer.

Crystal Mill
Colorado

Crystal Mill in Marble Colorado was built in 1892. As a power plant it provided compressed air for rock drills used in the 4000 foot Sheep Mountain Tunnel. The site of the mill is a natural waterfall. Building a wood dam across the river increased the fall and raised the water level but the dam washed out after a few years. The addition of a smaller dam upstream and a flume to carry the water to the penstock became a necessity. Placed on the bottom of a steel shaft fastened to the side of the penstock is a wooden waterwheel in a horizontal position. The "gear house" on the front of the building contained the mechanism for transferring the power to the large air compressor in the main building. The large timber foundation for the compressor is still in place. It is suspected that a large leather belt operated on a pulley at the top of the shaft to transmit the power generated. The waterwheel possibly also powered a stump mill. Additions include a new wood shingle roof and the log walls were strengthened. The mill is privately owned by the Crystal River Ranch Corporation.

In September 1984, the Aspen Historical Society and the Marble Historical Society decided the mill required emergency work to prevent the gear house from collapsing. Both groups had been raising money from public donations. Richard Haberman did the restoration. Wayne Brown and Oscar McCollum from Marble and Jack Roberts from Redstone volunteered to help. Supporting timber framework inside the building and one-inch cables were installed to shore up the gear house.

Story Courtesy of Oscar D. McCollum

Garrett's Mill

Connecticut

Garrett's Mill, located in the picturesque town of Bridgewater, Connecticut, has been a local landmark since 1796. Situated on Wewaka Brook, a tributary of the Housatonic River, it was restored in 1950 by Elmer Garrett, an engineer. Today, the mill is a working museum of hand papermaking. Mr. Garrett was a third generation papermaker who used the Bridgewater mill to demonstrate this lost art.

In the early days, the mill served the townspeople as a sawmill, a cider mill, ox-yoke mill and gristmill. The mill stands now, quaint and sturdy, known as one of Litchfield Counties' gems.

Today, the mill has a sixteen foot diameter overshot waterwheel installed by Mr. Garrett during the mill's restoration. A turbine replaced the original waterwheel in the early 1900s. In a rocky gorge, water pours over a twenty foot stone dam through the flume.

Machinery in the mill includes a Hollander style beater, vats, a Washington press, etching press, and many typefaces. In the large exhibition room samples of paper, papermaking molds, photographs of equipment used in papermaking and related items illustrate the history of papermaking.

The mill has sponsored papermaking workshops and provided a summer home for papermaking artists. Publications of the mill's Flume Press have included several volumes of poetry.

Story and photo courtesy of Florence Rome Garrett.
Photo by Brad Stanton.

ABBOTT'S MILL

DELAWARE

In 1802 Nathan Willey built and operated a gristmill near Milford, Delaware. Upon his death in 1812, the mill sold to James Owens and, later, Isaac Riggs. From 1821 to 1874, Johnson family members and others owned it. The core of the present building is believed to have been constructed during the later portion of the nineteenth century on the foundation of the original structure.

In 1919 miller Ainsworth Abbott purchased the mill. Abbott operated the mill until 1960, delivering flour weekly throughout the region. Abbott's Mill could produce ten to twelve barrels of flour per day. A barrel of flour weighed 195 pounds. The mill also shelled, cracked and graded corn, and produced ground livestock feed. Abbott used the "three-break method" of milling, a "break" being a stage in the reduction of grain to flour.

In 1963, the State of Delaware acquired Abbott's Mill. Renovations were done by the State Division of Historical and Cultural Affairs. Abbott's Mill was placed on the National Register of Historic Places in 1973, with the adjoining property added in 1979.

In 1981, the State leased the site to Delaware Nature Society and it became part of the Abbott's Mill Nature Center. A non-profit membership organization, the Nature Society fosters understanding, appreciation and enjoyment of the natural world through education, preserves ecologically significant areas and advocates the stewardship and conservation of natural resources.

Today, the preserved grain mill, 1905 house, stable and outbuilding stand at the northern end of Abbott's Pond. The machinery dates to the nineteenth century and includes both millstones and roller mills. A water turbine of iron later replaced the wooden waterwheel, probably of the overshot type that originally powered the mill. For more information call (302)422-0847.

Story and photo courtesy Linda Young.

Greenbank Mill

Delaware

The Greenbank Mill Historic District in Wilmington, Delaware represents three hundred years of milling along Red Clay Creek and reflects major events in United States history.

Milling began at Greenbank in 1677 when early settlers built the Swedes Mill. As the Middle Colonies earned the reputation as the "bread basket" of America, milling became the primary industry in Delaware. The merchant mill, constructed in the 1760s, ground flour for export to the Caribbean Islands and European markets. In 1790, Robert Philips purchased the mills at Sheriff's Sale. He contracted with his nearest neighbor, Oliver Evans, for automated milling equipment. During the War of 1812, Philips joined in speculation of Spanish merino sheep, known as "Merino Mania", which culminated in the construction of the Madison Factory. The factory, named for President Madison, contained cards, looms, jennies, a billy, a dye house, and lulling equipment.

The war ended and markets for American textiles declined but improved transportation systems resulted in a demand for wagon wheels and carriage parts. The Philips responded in 1850 by expanding with Thomas Blanchard's woodworking equipment and the first circular saws in New Castle County. They produced felloes, spokes, tool handles, and a patented line of camp furniture. By 1881, economic trends forced the sale of the mills. New owners relied solely on the grain operations that had supplied capital for the many other commercial endeavors. New roller mills were installed but the "bread basket" had now moved west and Greenbank became a custom feed mill. Tragically, arsonists burned the mill in 1969. Between 1969-1999, a group of dedicated volunteers worked to bring life back to Greenbank.

The restored merchant mill contains a reconstruction of Oliver Evans' revolutionary milling equipment. An eighteen-foot diameter oak waterwheel powers the mill equipment. The rebuilt Madison Factory houses the complex's history in exhibits and hands-on displays. A rotating model illustrates the many trades used to construct mills and milling equipment. Interpreters demonstrate 1820 and 1895 textile equipment. The dyehouse is an ongoing archaeological investigation and will be reconstructed when excavations are completed. The house and barn are being developed as museum areas with period rooms and agricultural exhibits. Heritage livestock graze the pasture surrounding the house and barn.

When completed, the Greenbank Mill will be a microcosm of early nineteenth century Delaware demonstrating daily life of a Quaker family--home industry, agricultural production of raw materials, and industrial life. Greenbank remains the only operating mill along Red Clay Creek where thirty mills once processed grains, wood, textiles, spices, iron and made milling equipment.

The thirteen-acre site with a merchant mill, woolen mill, 1794 miller's house, barn, pasture, and woodlands is an oasis of Delaware's agricultural and early industrial heritage all located in the middle of urban development. For more information call (302) 999-9001.

Story and photos courtesy Tony Shahan.
Photo by Mike Ciosek.

PEIRCE MILL

DISTRICT OF COLUMBIA

The Peirce family traced their origins back to Isaac Peirce, who left his Quaker parents in Pennsylvania, probably in the 1780s, to seek his fortune in Maryland. Peirce worked for Abner Cloud as a millwright, building and repairing mill machinery. In time, Peirce married Cloud's daughter, Betsy. In 1794 he bought a 150-acre tract that included a mill. By 1800, Peirce's holdings encompassed just over 2000 acres. In 1820, Peirce and his son, Abner (a stonemason), rebuilt the mill with blue granite quarried in the nearby Broad Branch area, giving us today's Peirce Mill. When Isaac Peirce died in 1841 he left the property to his son, who managed it until his death ten years later. The mill then went to Abner's nephew, Peirce Shoemaker, who owned it until 1891.

Neither the Peirces nor the Shoemakers were millers (one who runs a mill). Isaac was a millwright (one who builds mills) Abner, a stonemason, and Shoemaker, a jeweler. A succession of experienced millers operated the mill for them. Despite the Peirces' lack of direct involvement in milling, they were nevertheless interested in the technological changes taking place in the milling industry

For centuries mills had relied on the strength of the miller and his burly helpers. A major change in the industry occurred in 1795 when a young miller from Delaware, Oliver Evans, wrote a book, *The Young Mill-Wright's and Miller's Guide,* that revolutionized milling. The book made a science out of what had traditionally been a craft. With the system Evans designed, machinery did all of the work, from the time grain was poured into the receiving bin until dry, clean flour dropped into the holding bins. In 1820 the Peirces rebuilt the mill using this system.

Business usually remained brisk at Peirce Mill, especially during the 1860s. Often as many as twelve wagons a day arrived, loaded with wheat to be ground. A miller with a helper or two could grind more than seventy bushels a day on each set of buhrs, or millstones.

Today there are three pairs of millstones, all measuring about four and one half feet across and weighing about 2400 pounds. A miller could not properly operate the mill without the help of the stone dresser, who chipped away at the runner and bed stones with his pick-like millbill, sharpening furrows and shaping the raised areas about every three weeks. Finely ground flour prized for its baking qualities, made the stone dresser a valued employee. The most critical operation by the miller would be to adjust the distance between the stones. The runner stone, balanced on a spindle, revolved at about 125 revolutions per minute a fraction of an inch above the bed stone. An experienced miller could tell by rubbing flour between his fingers if the stones were properly positioned to produce the highest grade of flour possible. Hence the phrase " the rule of thumb". For more information call (202) 426-6908.

Story and photo courtesy of Dwight L. Madison, Supervisory Ranger, Rock Creek Park.

Old Munson Mill

Florida

In the 1800s, when the timber industry flourished in the Florida Panhandle, gristmills were scattered throughout the area. Cornmeal, rather than flour, was the staple product of the pioneer family's diet. They could eat it, sop it, or drink it. Without refrigeration, the meal retained its quality only for a short time. As a result, families ground only a small amount at a time. This meant they went regularly to the local gristmill to replenish their supply of cornmeal.

Not only was the gristmill a necessary part of pioneer Florida life, it became the center of the social activity, giving the hard working men and their families an opportunity to exchange information and pleasantries with their neighbors. It also gave the children time to play and swim in the lake. As times progressed, the gristmill's role in the community became less needed and eventually the gristmill disappeared entirely.

The gristmill at Munson Lake on the Blackwater River State Forest is powered by an overshot waterwheel. Water flows from above the wheel into the buckets causing the wheel to rotate and provide energy to turn the grinding stone. The water comes from two crystal clear springs to form the millpond, then flows through a wooden flume onto the waterwheel. This is the only overshot waterwheel mill in the state of Florida.

The bottom stone weighs 800 pounds and is stationary; the top stone weighs 1,200 pounds and is turned by the waterwheel to grind the corn. The kernels of corn are put into a hopper, which feeds onto the top stone. The miller may raise or lower the stone to make grits or cornmeal.

The Munson Gristmill provides a restful oasis. For future generations it offers a glimpse back in time when life may have been simpler. It stands as a tribute to a dedicated band of private citizens who donated their time and personal resources to preserve an integral element in the lives of Florida's pioneers.

Munson Gristmill staff give lectures about the Agrarian and Cracker culture. There is a wheelchair accessible boardwalk across swamps and wet lands connecting Krul Lake to Bear Lake. For more information call (850)626-4433 or (850) 957-6140.

Story and photo courtesy Warren Weekes.

Old Spanish Sugar Mill

Florida

The Old Spanish Sugar Mill in DeLeon Springs, Florida rests beside the spring named for legendary explorer Ponce DeLeon. Originally constructed to crush sugar cane, utilizing the power of sixteen to eighteen million gallons of water flowing from the spring daily, the mill boasts a one hundred foot undershot waterwheel. The actual construction date of the mill is a source of much debate. A chimney remaining on the site, part of the original sugar mill operation, displays a plaque proclaiming 1570 as the year of mill construction. Florida historians and archeologists disagree and feel evidence better supports early 1800s construction.

The mill has been destroyed at least twice, possibly three times, over the years. Again, historians are not in agreement. It is known the mill was destroyed during the Second Seminole War (1835-1842) and again during the Civil War when the mill ground corn for the confederate troops. Legend tells of millstones thrown into the spring although none have been recovered.

In 1961 the mill was again scheduled for destruction. Just an old building to many, Peter Schwarze felt quite differently. A fifth generation grist miller newly arrived in Florida, Peter saw a treasure that must be saved. He arranged to lease and restore the mill, moved his wife and daughter to the "Fountain of Youth" and began once again grinding flour in the mill building. He used his own design for the gristmill, an adjustable underrunner mill, with small french buhrstones, electrically powered.

The mill is operational and productive still and there is a restaurant on site operated by the Schwarze family. The mill building is now owned by the State of Florida and part of the Florida State Park System. Plans are currently underway to restore the waterwheel, which has not turned since 1976.

Story by Patty Schwarze.
Photo by J.C. Barger.

Shepard's Gristmill

Florida

The last Indian raid east of the Mississippi River occurred when a band of renegade Creek Indians massacred the McLane family a few miles south of Shepard's Mill on Telogia Creek. In the 1870s the area was unsettled unlike the vast areas of south Georgia just twenty-five miles to the north. In August of 1839, the Indians were captured and moved from the area so the land could be surveyed and sold to settlers by the Federal Government.

In 1855, the Federal Goverment deeded forty acres to the Pensacola and Georgia Railroad. This is the area where Shepard's Gristmill would later be built. At the time the railroads had an agreement with the government where they would receive free land for building a certain amount of track.

The Civil War took its toll of human life, livestock, and supplies of the few settlers in this area, and those families were now trying to put their lives back together and build a better future.

Much of the land was still unsettled in 1875, when an unknown builder constructed the first dam across Telogia Creek, and a one room, one story, waterpowered gristmill known as Shepard's Mill. He created an industrial and social center that became a landmark and an integral part of north Florida history.

As Shepard's Mill began to prosper, a sawmill and a cotton gin were added (between 1877 and 1894) to accommodate the needs of the area farmers.

In horse drawn wagons, a trip to Shepard's Mill became an all day affair and often the farmer's family accompanied him. While waiting their turn for the use of the mill, the adults would gather to socialize and discuss current politics and the business of farming. When the weather was good, the children would swim, fish, and play in the cool waters of the creek. Telogia Creek also witnessed many baptisms, as recent as 1990, and the high ground across the road from the mill was the site of the original Greensboro School. This one room log structure (built in 1896), served as the first area post office and hosted the area's Fourth of July celebration.

Shepard's Mill was the "heart" of the community and even today many older residents can recall their childhood "good times" and Shepard's Mill stories from their parents and grandparents.

In the early 1920s the Electrical Manufacturing and Developing Company purchased Shepard's Mill and attempted to generate electricity to sell to the community. Unfortunately, the project failed, but is a great example of industrial ingenuity.

In the 1930s, the original millstones (still on premises) were replaced, and in February of 1936, J.L. Swanson of Calhoun, Georgia installed a thirty-six horsepower turbine.

Other than these changes, Shepard's Mill remained basically the same from 1875 to 1960 when Wayne Hanna purchased the mill and began to modernize it in order to establish a "commercial" mill. Hanna was from a milling family who owned Hanna's Mill approximately three miles upstream on Telogia Creek and had grown up in the business. It was while Hanna owned Shepard's Mill that many changes occurred. The old wooden water box was replaced with a concrete and steel structure. Cleaner equipment and an elevator were installed in an outbuilding. A large grain bin, the two forty-two inch mills on the south side, roller bearings to the water turbine shaft, an elevator and separator system, a semi-automatic bagger, an electric driven mill and a second story were added to the building.

In 1969, after a twenty-three inch rain in twenty-four hours the spillway east of the mill washed out. It was replaced with a steel and concrete structure, built by Tony Miller and Adolph Whittle.

In 1978 Shepard's Mill closed and 103 years of industrial and social activity ended. Hanna had the unique distinction of operating the last "commercial" waterpowered mill in Florida at that time. The next ten years brought about different owners, but unfortunately nothing was done to preserve the mill, miller's house, or property. Although the buildings were deteriorated and the property overgrown, the area remained a favorite "fishing hole" for local residents, and in a small way continued to provide the "good times" for which it had always been known.

In July of 1988, after falling in love with Shepard's Mill and the area, Pat and Robert Dallas purchased and began restoring the mill, mill house, and property in an effort to save a part of history. After the purchase, the Dallas's discovered Shepard's Mill to be the last "commercial" waterpowered gristmill left standing in Florida, and only one other small, waterpowered family gristmill existed.

In 1989, Shepard's Mill earned the distinction of being Gadsden County's first historical site and structure in the support preservation. As a Florida Historic Site and Structure, Shepard's Mill became operable in the Summer of 1997. For more information call (850) 875-2694.

Story and photos courtesy Robert Dallas.

Butt's Mill
Georgia

Pine Mountain (Harris County), GA
Photo by J.C. Barger

Barker's Creek Mill

Georgia

For over 150 years the warm fresh aroma of cornmeal pouring from the grinding stones has been a weekly experience at Barker's Creek Mill. Barker's Creek Mill is a small community gristmill nestled at the head of Betty Creek Valley in the mountains of northeast Georgia, beneath the towering escarpment of the Nantahalla Mountains. The first family of white settlers originally built the mill in the 1830s. In the 1930s the floor of the mill gave way, destroying the grinding stones. In 1942 the Hopper family gave the derelict mill to the Hambidge Center, a handcraft-weaving center with property adjoining the mill. The structure, completely rebuilt on the old foundation, began operation again in 1944. Constructed of local field stone, pine and wormy chestnut, it first employed a primitive turbine referred to locally as a Willis Wheel. In 1948 it was replaced by a twelve foot waterwheel powering a Meadows Meal Master Mill and is the machinery still in use today.

The present structure saw many more years of daily operation, but the post war years brought gradual and dramatic changes to the valley. Shortly after World War II, textile mills came to the area, offering new opportunities for employment. In 1959 the completion of a paved road running the full length of the valley ended the isolated, self-sufficient life of the valley the mill supported.

In 1976 a drought finally brought an end to the steady but limited operation of the mill. It sat idle for several years until volunteers associated with the Hambidge Center began to open it for occasional demonstrations and grinding. Volunteers ran the mill sporadically until 1990, when Atlanta artist, Laurence Holden, revived regular operation.

Owned and operated by the Hambidge Center, it is listed on the National Register of Historic Places, and is the last one hundred percent waterpowered gristmill still operating regularly in the north Georgia mountains. In addition to the mill, the Hambidge Center has a gift shop and gallery. For more information call (706) 746-5718.

Story by Laurence Holden.

Barker's Creek Mill
Dillard (Rabun County), GA
Photo by J.C. Barger

Cresham's Mill
Georgia

Woodstock (Cherokee County), GA
Photo by J.C. Barger

Cresham's Mill
Woodstock (Cherokee County), GA
Photo by J.C. Barger

Cool Woods Mill

Georgia

Ooltewah, GA
Photo by J.C. Barger

English Mill
Georgia

Bird Crossroads (Macon County), GA
Photo by J.C. Barger

Fielder's Mill
Georgia

Fielder's Mill, in Talbot County, Georgia, is a waterpowered gristmill. Edlow Hart Fielder and his father, Lucious Pinkney Fielder, built the mill in 1930 on the old John Downs Mill site. Using horses, mules, and hand labor, the dam of the millpond was raised to give a headwater of fourteen feet with a pond twenty-five acres in size. The millpond is the headwaters of Patsiliga Creek. Patsiliga is a Creek Indian word meaning "pigeon roost", an appropriate name at the time because of the large numbers of carrier pigeons in the area.

The mill, powered by a Leffell type water turbine, produces about twenty-five horsepower. In 1965, Edlow Hart Fielder passed away and his widow, Ada Danley Fielder, continued the operation until 1967. Their grandson, Mike Buckner, then began operating the mill. At the time, Mike, fourteen years old, hired Charlie Spinks to drive and deliver the meal.

Through the mid 1960s the mill opened daily and more than fifteen area grocery stores were on a regular meal route. Up until the early 1980s customers brought their corn to be ground into meal or grits. The daily grinding done by Mike's grandparents is now done only by appointment and the mill route of fifteen plus grocery stores has dwindled to four.

Still a family operation, mill bags are printed on an old Chandler Price printing press. Even the children have their jobs cleaning the corn or wheat, stacking bags or boxes, or putting corncobs in the old potbelly stove when heat is needed.

Fielder's Mill has been a source of new stories several times over the years. The weather of the 1990s has been hard on the old mill. The March 1990 flood resulted in the dam breaking as more than twelve inches of rain was received within a twelve hour period. The flood of 1990 also washed away the dam. Hurricane Alberto almost washed away the mill building in 1994. In 1995, the Department of Natural Resources Historic Preservation program provided a grant to help restore the mill.

At the current fee of eight cents a pound or a toll of one fourth to grind someone's corn, "profits" are not the reason for continuing. "I like to see tradition continued." miller Mike Buckner says. "This is a tradition we certainly enjoy and one we hope our children will continue."

Fielder's Mill is the only remaining operational mill in the area. For more information call (706) 269-3630.

Story by Debbie Buckner.

Halls Mill
Georgia

Pine Mountain (Meriwether County), GA
Photo by J.C. Barger

Hamburg Mill
Georgia

The Hamburg Mill is located in Washington County, near Warthen, Georgia, on 225 acres of the Little Ogeechee River, which forms Hamburg Lake.

The history of Hamburg Mill began soon after the close of the Revolutionary War. In 1850, North Carolina native Richard Whathen, constructed a mill near the present site. He named the mill after the popular southeastern marketplace of Hamburg, South Carolina. The first wooden dam was constructed fifty to one hundred feet above the present dam site. Indians came from as far as Alabama to trade at the Georgetown trading post, about six miles east on the Ogeechee River.

T.B. Rachels and his brothers purchased the property around 1895. Oscar Harrison later acquired the property and sold to the Gilmore brothers, members of a prominent Washington County family. The Gilmores constructed the present dam and mill in the early 1920s. Construction took two years to complete.

The Gilmore brothers farmed much of the surrounding area. Capable of grinding both flour and corn, the mill also operated as a cotton gin and gristmill. The Gilmore brothers sold to the Hall family, whose members still live in the area.

The State of Georgia acquired the Hamburg property in 1968. During the active years of ginning and milling, the mill and adjacent store were community centers where socializing accompanied work related activities.

The Hamburg Mill is operated by the Georgia Department of Natural Resources, and is one of only two mills owned by the Georgia State Parks System. For more information please contact Hamburg State Park at (912) 552-2393.

Story and photos courtesy of Russell Hinson.

HEALAN'S MILL
GEORGIA

Healan's Mill rises out of the morning mist like a forgotten veteran. The mill has been known by many names, Head, Heard, Shore, Hyde and Turner are among these. It is currently owned by Mr. and Mrs. Fred V. Healan.

Millwright, Billy Head, built this old gristmill prior to the Civil War. Farmers came from miles around to have their corn and wheat ground, sometimes staying overnight. It has witnessed many historical events. During the Civil War Billy Head and his wife hid gold and silver in a hollowed-out log under the huge mill wheel.

Armor Rucker, serving with the Confederacy in Virginia when the war ended, walked the entire distance back to Georgia. Rucker arrived one day at the mill with a load of corn to be ground. When he decided to spend the night, his life changed forever. He met and fell in love with Nancy Florida Head, the daughter of Billy Head.

Bill Minor and his "gang" decided to rob the train at Browning Crossing, a short distance away. It is rumored they buried the money that night somewhere in the hills surrounding the mill. Many have searched the hills over the years but the money has never been found.

Around the turn of the century, other ventures began at the mill site; a wine press, gin, and sawmill.

Healans Mill still stands. Although silent now, it serves as a reminder of a time when you could hear the sound of water flowing over the wheel and wagons rattling and the scent of horses, harness and pipe tobacco filled the air.

Story and photo courtesy of Mrs. Fred Healan.

Hills Mill
Georgia

Ellaville (Schlex County), GA
Photo by J.C. Barger

LEE & GORDON'S MILL

GEORGIA

Three Gordon Brothers came to Chickamauga from Gwinnett County, Georgia, in 1836. They each bought several acres of adjoining land. James Gordon bought the mill site from Leroy Holiway and built the gristmill with attached store room. The mill served as the first general store in the area. Gordon managed the mill until 1857, when he turned it over to his son-in-law, James Lee. Lee passed the management of the mill to his eldest son Gordon Lee in 1889. Gordon Lee gave up the management of the mill to his younger brother Tom Lee. In 1929 Tom Lee sold the property to brothers Bill and Charley Wallace, who operated it until 1967. When Bill Wallace retired in 1967, he declared the mill retired too. He refused to sell or lease it. The mill stood empty and neglected from 1967 until 1993, when the Wallace heirs sold the mill property to Frank Pierce.

Mr. Pierce has rejuvenated the mill, repairing turbines and machinery. At the present time, grinding is being done on two original mills. Examples of all machinery used in the building operations have been placed on the main floor, cleaned and repaired to operational condition. The dam has been restored to original width and height and the original race and tailrace have been repaired to former condition.

The mill was occupied during the Civil War by both Union and Confederate armies. The Confederate forces under General Bragg were stationed at the mill the first few days of September, 1863. Later, during the Battle of Chickamauga, September 19 and 20, 1863, General Rosecrans stationed a Federal detachment at the mill to prevent Confederates from crossing Chickamauga Creek.

The general store room area houses many items of a by-gone time, including ceiling fans operated by waterpower. In addition to the mill there are two other buildings in the complex. The mill is listed on National Register of Historic Places.

For more information call (706) 375-6801.

Story and photo by Adair Brotherton.

Life College Mill
Georgia

Marietta (Cobb County), GA
Photo by J.C. Barger

Life College Mill
Marietta (Cobb County), GA
Photo by J.C. Barger

The Lindale Mill
Georgia

This old mill, built in the 1830s, rests on Silver Creek in Lindale. Captain Jacob H. Hoss owned the mill, first known as Hoss' Mill, for many years. Union troops partially destroyed the mill during the Civil War. Captain Hoss rebuilt the mill following the end of the war. It had a large metal wheel that carried the water from nearby Silver Creek. The bricks were made of clay excavated near the site. It operated as a gristmill for many years.

In 1895, Massachusetts Cotton Mills purchased the Old Mill and surrounding property and built a large textile mill across the street. For many years the mill stood idle.

During World War I the American Red Cross used the mill, loaned to them by the owners. Red Cross volunteers met weekly to roll bandages for the American wounded. In the early 1970s, The Lindale Garden Club undertook a restoration project. They leased the mill from West Point Pepperell for $1.00 a year. For several years the Old Mill housed Flower Shows and parties and many wedding receptions. During this time the waterwheel was replaced.

In 1986 Greenwood Mills purchased Lindale Manufacturing and The Old Mill. The Old Mill is listed on the National Register of Historic Places. For more information call (800) 444-1834.

Story courtesy Anne Culpepper.
Photo by Paul O'Mara.

The Lindale Mill
Cashiers (Floyd County), GA
Photo by J.C. Barger

Millmore Mill
Georgia

Sparta (Hancock County), GA
Photo by J.C. Barger

Nora Mill Granary

Georgia

Nora Mill Granary was built in 1876 on the banks of the Chattahoochee River in the mountains of northeast Georgia. The mill, a large four story building, has a one hundred foot wooden raceway powered with a turbine.

On the same site is evidence that a mill stood for over 200 years, but John Martin, a Scotsman, built the mill that stands today. Martin came to Georgia to mine for gold, and unlike most miners, made Nacoochee Valley his permanent home.

In 1905, Lamartine Hardman, a former governor of Georgia, bought the mill and named it "Nora Mill," in memory of his sister. Nora Mill remained in the Hardman family until 1998, when Tom Slick, with Nacoochee Village Ltd., purchased it and 300 surrounding acres. A succession of millers came and went through the years. Ron Fain is now the miller.

The Fain family roots began just a few miles north of Nora Mill, where Ron's Father, George, resided until 1917 before moving to Ohio in search of work. While in Ohio, George married Florence McPherson. The McPherson's raised four children. Ron, the youngest boy, grew up in Akron and married Rita Lepera. After graduating from the University of Akron and retiring from the Army as a Lt. Colonel in 1979, Ron and Rita, and their five children returned and settled in the Fain family homestead.

As Ron got involved with Nora Mill, he became so interested in preserving this bit of history, that he buried himself in books and publications about waterpowered mills. He also sought help from knowledgeable individuals, including Clyde Keltner of Tennessee and long time miller of Nora Mill, Tom Farmer. Now well educated in the lost art of grain milling by waterpower, Ron stays active in The Society of the Preservation of Old Mills.

Today at Nora Mill, the original, 1500 pound, forty-eight inch stones that labored in 1876 still turn and grind grains. Nora's Mill is an operational mill with a gift store. For more information call (800) 927-2375.

Story courtesy of Joann Fain Tarpley.
Photo courtesy Travis Massey, Magic Craft Studios.

The Old Mill
Georgia

The Old Mill, constructed in 1930, in Rome, Georgia is one of the most popular spots on the Berry College Campus. Its wooden overshot waterwheel measures forty-two feet in diameter and is considered one of the largest in the world. The wheel's iron hub was originally used at Hermitage, an early manufacturing community near Rome. The hub was given to Berry by the Republic Mining and Manufacturing Company. Automobile magnate Henry Ford had the hub moved to Berry, where Berry Students rebuilt the wheel.

The mill was operated for many years by Green Berry Goodson, a white-bearded miller from the mountains. Those who came to see the picturesque mill did not easily impress Mr. Goodson. One day Martha Berry arrived in her car with Mrs. Sarah Delano Roosevelt (mother of Franklin D. Roosevelt, President at that time) and Mrs. Thomas A. Edison. Martha instroduced both to Mr. Goodson. Although Mr. Goodson barely acknowledged Mrs. Roosevelt, when Mrs. Edison was introduced he said, "Wal, I guess you're worth coming out to shake hands with."

In 1977 the wheel was completely rebuilt as a cooperative project involving Berry students, staff, alumni and friends. It was dedicated to the memory of Mr. Gordon Keown on June 3, 1978. A 1905 graduate of Berry, Mr. Keown served as acting director of the Berry Schools from 1942-1944. Perhaps his most lasting contribution to Berry is the more than 33,000 acres of land he acquired for the school.

In recent years, staff and student volunteers have restored the Old Mill's mechanism and made the grinding of cornmeal possible once again. Berry College campus takes you back in history with the Old Mill, an antebellum home, museum, carriage house, gift shop and gardens. Historical photographs of the Old Mill are displayed in the museum. For more information call 1-800-220-5504.

Story courtesy of Berry College.
Photo by J.C. Barger

Old Parrish Pond Mill

Georgia

Old Parrish Pond Mill, built in 1880, provided valuable service to the local community for many years. The mill began as a sawmill and eventually became a gristmill. Local townspeople would bring corn to be finely ground. It is also a covered bridge and one of only a few in the state today. The turbine, waterpowered and controlled by floodgates, could be lifted when necessary. The covered bridge had a public access road through it, but eventually closed because of the wear and tear on the mill. The mill is one of the many highlights of George L. Smith II State Park, named after one of Georgia's great legislators.

On December 9, 1998 a grand opening was held at George L. Smith for the gristmill. The mill, dormant for thirty-five years, is once again operational. A new turbine has been installed and replacement of the siding and outside windows are in process. For more information call (912) 763-2759.

Story and photo courtesy of Robert Dowse.

Prater's Mill
Georgia

Once the home of the Cherokee Indians and a campsite for both Confederate and Union armies, Prater's Mill and the surrounding countryside along the Coahulla Creek are rich in southern tradition.

Built and developed by Benjamin Franklin Prater in 1855, the mill and hundreds of acres of farmland were used for the production of cotton, wool, wheat, corn and oats. The mill processed grains for the neighboring farmers into cornmeal, flour and feed for animals. Constructed of hand-hewn heart pine timbers with mortised and pegged joints, the ruddy three-story mill operated with waterpower by two Leffel-type turbines and one Davis turbine, generating seventy-five horsepower. Hand-carved augers and bucket elevators carried the grain through multiple levels of cleaning, grinding and sifting equipment. There were three stands of roller mills for flour production.

During the Civil War, the mill served as a campsite for soldiers of both armies. On February 23, 1864, six hundred Federal Cavalrymen, commanded by Colonel Eli Long, based there after a skirmish in Dalton. On April 13, 1864, twenty-five hundred Confederate cavalrymen, led by General Joseph Wheeler, camped at Prater's Mill on their way to Tunnel Hill after a confrontation in Murray County. Fortunately, nothing was destroyed during either army's visit.

By 1879 there were at least thirty-two gristmills in Whitfield County. Prater's Mill alone remains, and is also one of the few still operational in the State of Georgia.

The Prater family operated the mill until the 1950s. It sold to the Church of God of the Union Assembly, who ran it until the 1960s. The buildings and machinery fell into serious disrepair. In 1971, the Prater's Mill Foundation (all volunteers) began extensive restoration and preservation efforts. Prater's Mill is now listed on the National Register of Historic Places.

In 1973 the mill sustained heavy damage by the Great Chattanooga flood, with five feet of water inside the building. Then in February, 1990, high waters again attacked the mill, washing away a thirty foot load-bearing heart-pine beam causing other destruction.

In March of 1994, the old mill was again damaged by a flash flood. On May 20,1995 arsonists set fire to the mill, which sustained $81,000 in damage.

As well as the fascinating intricacies of the equipment at the mill, a beam remains bearing "B. F. Prater's" signature. The names and dates of former customers and millers are also scrawled on the walls and steps. The Prater's Mill volunteers aptly refer to the writings as "genealogical graffiti". For more information call (706) 694-6455.

Story and photo courtesy Prater's Mill Foundation.

Prater's Mill
Dalton (Whitfield County), GA
Photo by J.C. Barger

Peeler's Mill
Georgia

Flintstone (Walker County), GA
Photo by J.C. Barger

Ragsdale Mill (Cantrell's Mill)
Georgia

Homer (Banks County), GA
Courtesy Society for the Preservation of Old Mills

Skeenah Creek Mill

Georgia

Morganton (Fannin County), GA
Photo by J.C. Barger

Skeenah Creek Mill
Morganton (Fannin County), GA
Photo by J.C. Barger

Short's Mill
Georgia

Clarksville Vicinity (Habersham County), GA
Photo by J.C. Barger

Starrs Mill
Georgia

Fayetteville (Fayette County), GA
Photo by J.C. Barger

Stone Mountain Gristmill

Georgia

DeKalb County, Gerogia
Photo by Palmer Werner

Stone Mountain Gristmill
Stone Mountain (De Kalb County), GA
Photo by J.C. Barger

Sylvan Falls Mill

Georgia

Sylvan Falls Mill has operated as a gristmill for over 150 years. Construction of the mill took place in 1840. The mill, of wormy chestnut wood, is built on a solid foundation of rock that also serves as flooring in some parts of the mill.

In 1946, the new owner, R. N. Miller, replaced the original wood waterwheel with a twenty-seven foot, 10,000 pound steel wheel he moved from Tennessee. The waterwheel is one of the largest in the United States.

In 1977, Maynard and Florence Murray purchased the mill, and after renovations converted it to a permanent residence.

Sylvan Falls Mill sits in the scenic Appalachian Mountains of northeast Georgia and now operates as a bed and breakfast. For more information call (706) 746-7138.

Story and photos courtesy of Deborah O'Brien.

Sylvan Falls Mill
Dillard (Rabun County), GA
Photo by J.C. Barger

Tabb's Mill
Georgia

Situated on a 100-acre cypress pond, Tabb's Mill is one of the few waterpowered gristmills operating in the State of Georgia. Located northwest of Newton, Georgia in the Patmos Community, Tabb's Mill has also been known as Ivey's Mill, established in 1838 over Mill Creek, a tributary of the Ichauway-Nochaway Creek.

R.D. Ivey, Sr., born 1811, died 1871, was the original owner. After his death, his wife inherited the mill. In 1892, Mrs. Ivey deeded the mill property to two of her sons. The stipulation in this transaction read, "The right and privilege of getting my supply of bread without cost during my natural life is reserved in this gift." Members of the Ivey family retained ownership for over eighty years,

A waterpowered turbine turns the large rotating millstones, which grind corn into meal. The water furnished by the pond powered a sawmill, cotton gin, and the gristmill. Slaves built the earth part of the dam with ox carts and wheelbarrows used to transfer the dirt. Later, that section of the dam became a public road and was designated as a stagecoach route, enacted by the Georgia General Assembly.

A covered bridge was once attached to the west side of the original mill. Spillway gates regulate the desired water level, thereby draining the pond if necessary for repair work.

People traveled many miles to Baker County to have their corn ground into meal. Around 1900, the mill operated full-time, six days a week. Early day patrons came by wagon and often camped on Mill Creek for the night. At maximum capacity, approximately forty bushels of shelled corn could be ground in a day from a single millstone.

Fire destroyed the mill around 1895. Determined to rebuild, the Ivey family erected a second mill.

In 1925, the Ivey family sold the mill. By 1938, the mill again needed repair. At this time, new owners restored the mill. Lester W. Tabb purchased the mill in 1954. At present, the mill route has ceased. The mill is operational, but is used today on a much smaller scale in comparison to the wagon trails of years gone by.

Through the years, the mill has been a popular fishing paradise and favorite swimming hole. Below the mill baptisms have taken place and political forums held. Trips to the mill were usually fun for young and old because of the fascination of the large stones that turned as grains of corn were ground into fine meal. Many times the mill operated from sunrise to sunset.

Tabb's Mill is the only waterpowered gristmill still in use in Baker County. The mill has withstood the elements of time and the foundation is still solid in spite of the floodwaters of the 1900s. For more information call (912) 734-4143.

Story and photo courtesy Sandy Tabb Hodges.

Wilkerson Mill
Georgia

Wilkerson Mill
Palmetto (Fulton County), GA
Photo by J.C. Barger

Fabyan Windmill

Illinois

Two German craftsmen originally built this sixty-eight foot Dutch Windmill on another site, but it now stands in the Fabyan Forest Preserve, Geneva, Illinois. Moved to Geneva and reassembled October 15, 1914, it took nineteen months to reconstruct. The mill is estimated to be more than 130 years old. Colonel George Fabyan purchased it for $8,000.00. Transportation and rebuilding costs exceeded $70,000.00. Fabyan agreed to have a Danish millwright, Mr. Rasmussen, oversee the reconstruction. John Johnson, from Batavia, and the Wilson Brothers Construction Company assisted.

Roman numeral markings carved into the beams used in reconstruction are still visible. It took thirty-three workers mixing concrete by hand to build the foundation. The foundation is forty-two inches deep and twenty-six inches thick. Inside the mill the beams and shafts are of cypress wood and trimmed with black walnut. New gears had to be made of wood.

While placing new shingles on the mill, a paper was found under the shingles signed by men who worked on the mill in 1914-1915. They were Oscar Wilson, Charles Lindahi, Ben Anderson, Carl Peters, Martin Seastrom, and Dick Swanberg.

In 1973, the Geneva Chronicle interviewed William Benjamin Munson, a member of the first crew to work on the Fabyan windmill and employee of the Egar E. Belding Co. of West Chicago. The Belding Company was employed to move the structure once it had been disassembled. The transporting crew, with the use of a team of mules and a twenty foot extending wagon, moved the fifty foot uprights to the building site.

At the top of the mill is a huge cogged wheel turned by the wings and wind, which then rotates a rod rimming the height of the mill. This turns two grinders and a sharpening wheel in the basement. The cogs are of wood, carefully fitted, as gears must be to do a mammoth job. Throughout, dowels glued into drilled holes hold the mill. There is scarcely a nail to be seen.

During World War I, the mill served its purpose by grinding flour. Paper bags to hold the flour are still in the basement, together with an old tool box and hand drill. The huge antique oven was used to bake bread, although this was short-lived as the flue system was inadequate. Several honors have been accorded to this windmill. It is listed on the National Register of Historic Places, and honored in 1980 by its selection to be on a U.S. Postal stamp. This fifteen-cent stamp was part of a series of five American windmills included in a stamp booklet called WINDMILLS USA. For more information call (708) 232-4811.

Story and photo courtesy Jon Duerr.

Franklin Creek Gristmill
Illinois

History is much more than facts and figures. In the old mills are historical architecture consisting of massive timbers with incredible joinery. It is natural to wonder what the people were like and how they performed such impressive acts of craftsmanship. Imagination must be used to relate to the mill builders and operators. It is a challenge to accurately, mentally recreate the characters of those living a century and a half ago.

The recently reconstructed Franklin Creek Gristmill in Lee County, Illinois illustrates nineteenth century heavy-timber type construction, and allows the experience of meeting mill builders and operators. Volunteers raised hundreds of thousands of dollars and performed thousands of hours of labor to reconstruct the Franklin Creek Gristmill. Technology has modernized our world, but the mill is evidence that the work ethics and community spirits of rural America are alive and well.

Reverend Joseph Emmert and his son-in-law Christian Lahman came from Franklin County, Maryland in 1843 looking for natural resources suitable for construction and operation of a gristmill. The success would be measured by the ability to combine hard work with the creative use of abundant resources such as water, wood, soil and even air. Surprisingly, the Emmerts and Lahmans selected a mill site a half-mile from the water source. Their project required excavating a long raceway and a two and one half acre millpond. Horse power and hand shovels moved a tremendous amount of earth. Early builders chose the site because of its accessibility and elimination of flood and ice damage.

Still, reconstructing a mill is an overwhelming task. Success is still measured by hard work and determination. Mills are built and operated by special, gifted people. The Franklin Creek Preservation Area Committee and friends are the group who met this challenge. The Committee demonstrated their resolve initially by developing and operating the Franklin Creek State Natural Area on a first ever volunteer basis. In 1987, President Reagan invited the Committee to the White House as a recipient of his "Take Pride in America" award in its first year. The Gristmill reconstruction followed the state natural area improvement work as a facility to preserve and interpret the area's cultural heritage. For more information call Franklin Creek State Natural Area (815) 456-2878.

Story and photo courtesy of Elmer Stauffer.

Graue Mill
Illinois

Frederick Graue built Graue Mill and three successive generations owned and operated it. The mill ground wheat, corn, oats and buckwheat and is still remembered vividly by older local residents as a leading economic unit in the community.

Nicholas Torode and Sherman King, who came to this area the year before, built the original dam at this site in 1837. The waters powered the sawmill but the dam washed out a few years later. Around 1844 a log dam replaced the damaged one. At that time they also built a log mill, which burned in 1848.

Fred Graue, listed in Richmond's history as one of the settlers prior to 1835 in York Township, was born in Hanover, Germany, January 1819. He and William Asche purchased the mill site following destruction of the log mill by fire. They erected a sawmill, and after three years Graue bought the entire interest. By the summer of 1852 he had completed the brick mill, a building forty-five feet by twenty-eight feet, three stories high, with a basement. There were two runs of buhrs and according to Blanchard's history, "has since run same mostly on custom grinding. He has about 200 acres of land which he carries on, but gives his attention personally to his milling."

A millwright brought from New York installed the mill machinery. None of the original plans of the building are on record, although the Historic American Building Survey has developed accurate plans of the structure as it stood in 1934.

Clay taken from Mr. Graue's farm and burned in a kiln near the site made the brick used for the building. The stone for foundations and trim came from a limestone quarry. The timbers for posts, girders and joists are of white oak.

The millrace ran east from the south side of the millpond to the wheel (later a turbine) and emptied the water under an arched opening at the lower end of the race, the water joined the stream well below the dam. Just west of the mill were gates in the race, regulated by the flow of water past the wheel, to relieve the flow over the dam. At such times the wheel was allowed to idle.

A crib and plank dam replaced the log dam in 1879. Some time before 1874 the waterpower furnished by Salt Creek became supplemented by steam power, a boiler and stack were built on the opposite side of the race. Records show the installation of a fifty-six inch Standard Leffel Upright Turbine in 1868.

During the Civil War cane syrup made in this mill would be boiled in steam. In 1893 a hydraulic cider press was installed in the mill.

In 1916 the crib dam built in the 1870s washed out. No dam existed at this site until the new concrete and stone structure was constructed eighteen years later.

From 1916 to 1924 the mill operated occasionally. The miller foreman for many years, Patrick Kammeyer, died in 1928. F. O. Butler acquired the property in 1921, and in 1931 the mill became part of the properties of the Dupage County Forest Preserve District, forming an addition to the Fullersburg Forest Preserve tract, Oak Brook, which lies along Salt Creek. Story and photo courtesy of James Blaha. For more information call (603) 655-2070.

Hiram Buttrick Sawmill

Illinois

Antioch (Lake County), IL
Photo by N. Votava
Courtesy Society for the Preservation of Old Mills

Morrison's Old Mill

Illinois

Annan's gristmill at Morrison has shared in the pages of the Whiteside County history since the construction of the mill in 1859. Known first as Annan's Gristmill and later as the Unionville Roller Mill, area residents simply refer to the building as the "old mill."

William Annan and John A. Robertson built the original mill on the north shore of Rock Creek, which served as a water source to supply the power to operate the mill. A frame dam structure replaced the original dam of brush construction in 1862.

Around 1867 Anna became sole owner of the gristmill. He and his son operated the mill for more than forty years. The Annan heirs sold the mill property to Orville Hawkins in 1926.

According to Wayne Bastion's *"A History of Whiteside County,"* the old mill building measured thirty-two feet by forty-two feet and featured a basement, two stories and an attic area. There are three runs (pairs) of large stones, forty-eight inches in diameter. Before 1877, a run of buhrs in the so-called pony size, thirty-two inches in diameter, were added. Prior to 1885 two sets of rollers were added.

The grey stone used in the old mill structure may have been taken from a local quarry.

Historians indicate that the old mill did custom and commercial milling in it's earlier days of operation. The Daily Gazette, Sterling-Rock Falls, Illinois, December 24, 1976.

The above photograph depicts the mill as it stood in 1858. Current status is not known.

Courtesy Society for the Preservation of Old Mills.

Prairie Mill

Illinois

The Prairie Mill is one of the first structures seen when entering Golden from the south. It was built on a site purchased by Henry R. Emminga in June of 1872. He bought thirty-three acres of land adjoining the south line of Keokuk Junction, from Col. William Hanna. In August of the same year, carpenters under the direction of H.R. Emminga, began work and completed the first story September 2, 1872. The following winter and spring Mr. Emminga made most of the wooden machinery used in the mill. The main drive of the master wheel, cam and sprocket type, were twelve feet in diameter and made entirely of hard maple. It required almost eight months to finish. In the fifty-two years of continuous service this wheel showed very little wear. The main shaft on which the wheel and four fans were fastened, was made of cast iron and weighed 4,700 pounds. The bearing in which it rested weighed 340 pounds. These were made by special order by the City Foundry in Quincy. Emminga drew all the plans and made the patterns for the castings.

There were two sets of lava buhrs ready for use in September 1, 1873 and the grinding of mill foods began. In August of 1874 the third set of buhrs was in place and they measured five feet in diameter and weighed 5,000 pounds. With the three sets of buhrs (millstones) it had a capacity of 500 bushels of grain a day. Thirty-five loads of rock were hauled from a creek seven miles away by farmers of the community. The large elm, oak and hard maple timbers were sawed by Mr. Buss near Mt. Sterling.

The mill was used in the manufacturing of buckwheat flour, rye and graham flour, corn meal, and ordinary mill feeds. The mill held the highest reputation on its products and its fame spread over the United States and many foreign countries.

In the fall of 1878, H. H. Emminga bought his father's interest in the Mill and remained sole owner and proprietor until his death, December 1915. It was then taken over by the Consolidated Cereal Company, with J.J. Emminga and F.B. Franzen consolidating this windmill with the custom mill east of town.

A strong wind tore off two of the four fans on the mill in 1924. Mr. Franzen installed a thirty horsepower gasoline engine, assisted by his son, Ralph, and son-in-law, Walter Reynolds, and continued the work for some time in this manner.

The windmill is now being restored by the Golden Historical Society.

Story courtesy of the Golden Historical Society.

Adams Mill
Indiana

John Adams moved from Pennsylvania to Indiana to build a mill. Walking along Wildcat Creek, he looked for the best place to settle. On the North Fork, just on the edge of an oxbow, he could make a dam and cut a millrace and tailrace to run an undershot waterwheel. In 1835, he first built a sawmill. In 1845 he erected a new building four stories high and water turbines were used for power. He had two buhrs for wheat and two for corn.

This was the era of the stagecoach. In 1850 the Federal Government used Adams Mill for mail for fourteen years. In 1863, Mrs. Rankin acted as postmaster and taught school on the third floor for one year.

Warren Adams took over the mill in 1858 and in 1864 the Wildcat Masonic Lodge 311 began and met on the third floor of the mill for four years. In 1887, Levi Bishop, Warren's son-in-law took over the mill. He found that rollers proved cleaner and faster. Using rollers the small turbine could power the full production of flour.

Jessie Johnson, Bishop's son-in-law, took over the mill in 1900. He used the large turbine to generate electricity. He and his brother ran Wildcat Utility from 1900 to 1917, with time out for the 1913 flood. From 1918-1929, Mr. Badgley, unrelated to the family, took over the mill. Mr. Hahn owned the mill from 1929-1935. He also owned a mill in Burlington. From 1935-1942, Mr. Rueben and Mr. Lett had the mill. Lett's son-in-law, Mr. Grayson, operated it. The mill needed restoration work.

Claude Sheets owned Adams Mill from 1942-1974. He restored siding, windows, and machinery. His brother, Dewey Sheets, came from the Navy to run the mill. By 1952 changes in health food laws made them decide to convert the mill to a museum of Americana and illustrate the old style of grinding. Antiques filled the mill, from a Model-T Ford to coffee grinders.

In 1974 James Broadhurst bought the mill to use as a tourist attraction and limited the antique collection to the 1800s. The mill was recognized by Historic Preservation.

Jill and Mark Scharer bought Adams Mill in 1992. The Friends of Adams Mill Valley Inc. lease the mill. This volunteer organization uses the mill for family oriented history, education and recreation.

Story courtesy of Barbara Cottingham, President of Friends of Adams Mill Valley Inc.

Beck's Mill

Indiana

Beck's Mill is located in Washington County, Indiana. The millhouse, built in 1864, is the third mill on the site. George Beck and his sons built the two and one-half story, front gabled millhouse. A rough cut limestone foundation supports the structure.

The mill still has its original horizontal boards of weathered white poplar clapboard. It used a turbine and a waterwheel in its operation. In 1897 a sixteen-foot iron overshot waterwheel replaced the existing wooden one. Power for the flour mill was supplied by a small turbine.

The first mill on the site was constructed by the Beck family in 1808, the second in 1825 and the third in 1864.

Usually the mill operated twenty-four hours a day from 1865-1890. By 1901, Beck's Mill discontinued grinding flour but continued to grind feed and cornmeal until some time in the 1960s. The grindstone, other original mill equipment and machinery is intact.

Beck's Mill is one of five properties located in Washington County listed in The National Register of Historic Places. Information on the photo shown dates it sometime in 1998-1999.

Courtesy Society for the Preservation of Old Mills.

Bonneyville Mill
Indiana

Bonneyville Mill is the oldest continuously operating gristmill in Indiana, producing quality flour for over 150 years.

During the 1830s, Edward Bonney had a vision of a rapidly growing city centered around a sawmill and a gristmill on the Little Elkhart River. He selected the location due to its proximity to the Toledo/Chicago trail and the navigable St. Joseph River, which provided access to Lake Michigan commerce. When the railroad bypassed the Bonneyville site and the proposed canal system did not develop, Bonney's dream of an urban hub ended. Bonney sold the mill and went into the tavern business. Later accused of counterfeiting, he fled the area as an outlaw.

Bonney installed a horizontal waterwheel and turbine, which could produce larger quantities of horsepower than the traditional vertical waterwheel. This innovation enabled Bonneyville Mill to function as a progressive mill of the era. During the 1880s, milling advanced rapidly using new innovations to speed the production of flour. Many mills replaced grist stones with the recently invented roller mill to grind flour. Bonneyville Mill chose to remain a "progressive" mill of the Civil War era, committed to serving the local farmeres and a few merchants.

Bonneyville Mill owners met the challenge of the Depression by forming Bonneyville Power and Electric Company in 1932, supplying hydroelectric power to about forty-five customers. The mill produced livestock feed and offered custom grinding. Speculating in popcorn, dealing in fertilizer, and grinding flour became a way of life into the early 1950s.

Advancing technology and regulatory legislation eventually overwhelmed the county milling business, and the mill owners were forced to sell out.

Elkhart County Chapter of Michiana Watershed, Inc. purchased the mill in 1968 and presented it to the Elkhart County Park and Recreation Board in 1969. It remains an example of workable antiquity. The mill grinds corn, wheat rye and buckwheat the old fashioned way, slowly, methodically as if time were only water over the dam.

For more information call the Elkhart County Park and Recreation Department @ (219) 535-6458.

Story courtesy of John W. Jenney, Mill Manager.
Photo courtesy of Jeff Johnson.

Greenfield Mills

Indiana

In 1835 Samuel Burnside built a sawmill on the banks of Crooked Creek (now the Fawn River). A few years later Peter Beisel began construction of a gristmill nearby. This mill was put together with square nails and wooden pegs. On the third floor, five fifty-foot hand hewn white oak beams support the roof. Many of the floor supports and sheeting boards were cut at the old sawmill from yellow poplar (tulip) trees. Some of these boards measure over thirty-two inches in width. Beisel ran into financial trouble and was unable to complete the mill. Amos Davis completed it in 1846. In the following years the mill went through several different owners until purchased by Henry Rinkel in 1904. At that time, the mill was being used as a dance hall. The dam washed out months before.

Many neighbors thought Rinkel was crazy for trying to make a go of the "jinxed" old mill. Still, they came with their picks and shovels to help rebuild the dam two different times. It took two long hard years to fix up the old mill. The previous owner left only two sets of stone buhrs, an obsolete reel for sifting, and two waterwheels in very poor condition.

The old waterwheels were removed and Rinkel built a new flume and installed three vertical turbines. The first turbine drove three stands of rolls for grinding white flour and the second drove one set of stone buhrs for grinding whole wheat flour, corn meal, and buckwheat flour. The third turbine drove the second set of stone buhrs for grinding feed. He also installed twelve bucket elevators to move the flour through the milling process and added a sifter.

Finally in 1906, Henry Rinkel began producing flour. Over the last ninety years the spelling and the machinery have been changed but the same quality flour is still produced.

In 1925, Henry put up a brick building and installed a seventy-two kilowatt hydroelectric generator. His son George returned to Greenfield two years later and the business became H.G. Rinkel & Son. Between 1932 and 1934, they added an eighty kilowatt generator. A year later they built another addition connecting the original building and the brick building. In this area they installed a complete buckwheat mill.

In 1949, four years after Henry's death, George's son Howard completed his milling degree and returned to the mill. Over the years they continued to upgrade the mill. Much of the equipment was purchased from small mills that had gone out of business. Sadly, while there were once hundreds of small wheat mills in Indiana, only a handful are left in the country. Greenfield Mills is the oldest commercial waterpowered mill left in the state. It is also one of Indiana's smallest electric utility companies, serving two mills and eleven homes.

In 1974 George, Howard and Howard's son, David, broke ground on a "New Mill" building and grain elevator down the road from the "Old Mill". In this building they installed an Italian Pneumatic Roncaglia Mill, which moves everything by air. This "New Mill" produces the same or slightly more flour, in one large "noisy" room, than the "Old Mill" produces on four floors. George Rinkel lived long enough to see this mill run but full production started in 1981, three years after his death.

For many years the mill ran full time producing flour used to bread Kentucky Fried Chicken. However in the late 1980s, after Colonel Sander's death, the recipe changed and the company no longer used the same soft winter wheat flour. In 1990 an Ohio company approached Howard and his son Dan about grinding organic buckwheat flour. Instead of buckwheat they began grinding both organic whole wheat and white pastry flour. Howard's daughter, Joyce, then returned to help with running the "New Mill" while Dan continued to run the "Old Mill".

The present owners of the mill are Howard and Helen Rinkel. Third generation Howard still puts in a full week's work and Helen is the secretary treasurer. Fourth generation Dan and his sister Joyce complete the full time employee list. Their brother Dave and fifth generation grandchildren; James, Amanda and Jordan can often be found pitching in during busy times. Greenfield Mills is truly a family business. For more information call (219) 367-2394.

Story and photo courtesy of Helen Rinkel.

Mansfield Roller Mill

Indiana

Since 1819, a mill has graced the banks of Big Raccoon Creek in the village that would become Mansfield in Parke County. James Kelsey and Francis Dickson built the first mill, a simple thirty feet by thirty feet log building. According to legend, glacial stones from a nearby farm were used for grinding. From that simple beginning, the Mansfield Mill came to represent an important part of Parke County history.

The village of Mansfield quickly grew up around that early mill. To accommodate the needs of the early settlers, a sash saw mill and a carding mill were added to the existing gristmill. A general store, blacksmith, cooperage, and miscellaneous businesses were built near the mill, and eventually a church and school were organized. Like many locations, without a mill there would have been no village.

In 1875, Jacob Rohm bought the mill property from James Murphy. Rohm came from an old milling family in Pennsylvania, but had moved around the Midwest while building and operating mills. At the time, the Mansfield Mill was a three-story building with three runs of stone operated by a turbine. The Rohm family lived in the mill until a house could be secured.

By 1880, Rohm decided the building did not suit his needs and built a new mill north of the existing building, measuring twenty-four feet by thirty-six feet. It is the original section of the present mill. The conversion from stone grinding to roller system in 1886 required another section be added. By 1893, Rohm's sons, George W. and E.H. Calvin Rohm, owned the business and added the third story to accommodate more equipment.

The Rohm family, quite successful in the milling business, eventually owned five mills, three grain elevators, sawmills and other enterprises in Parke County. George W. Rohm served on the War Production Board during W.W.I and as president of the Indiana Millers Association. Milling had become the largest industry in Parke County and the Rohms were considered the wealthiest family.

In the 1920s financial difficulties plagued the mill. Rohm Brothers and Company eventually went into receivership and the Mansfield Mill sold at Sheriff's Auction for back taxes in 1929. The Reeves family purchased and operated it as a feed mill until 1968.

The mill had various owners from the 1970s to the 1980s. The Hutcheson and Dalton families donated the Mansfield Roller Mill to the State of Indiana in 1995. Now part of the Indiana State Museum System, it is considered the finest example of early roller milling in Indiana. All three floors of machinery are intact and a working turbine remains. The mill is still operated occasionally. For more information call (765) 344-0741.

Story and photo courtesy Micki Morahn.

Metamora Gristmill

Indiana

Nestled between picturesque hills in the heart of the Whitewater River Valley, Indiana, is the quaint community of Metamora. Here the Metamora Gristmill sits beside the Whitewater Canal. Metamora flourished as a thriving port with the development of the canal during the early 1840s.

Jonathan Banes, a carpenter, began the Metamora Cotton Factory, the forerunner to the mill. It functioned as a cotton mill for eleven years, utilizing a thousand spindles. Unfortunately, financial difficulties required Banes to convert the business to a flouring mill in 1856, at which time he took on a partner named Murry.

Just a year later they sold the business to John Curry and Son. It was sold again in 1863 to Thomas Tague, who operated the mills under the name Hoosier Mills. William McClure purchased it in 1877 and renamed it Crescent Mills.

Although ownership frequently changed, the mill remained an integral and vital part of the community over the years.

Regrettably, the original three-story frame building burned in 1899. It was promptly rebuilt in 1900 and converted to a two-story brick building. A "tub" wheel in the basement was originally fed by water from the canal via a wooden flume.

With the demise of the canal in the early 1860s, a turbine placed in the canal lock adjacent to the mill, turned faster and more efficiently than the wooden waterwheel. The water dropped over a "lip" constructed on the upstream end of the lock and fell onto the turbine.

Since 1946, the State of Indiana has maintained a fourteen mile section of the canal, including the gristmill, which has been designated as a National Historic District.

In 1973 restoration of the mill included installation of a twelve foot wooden breast wheel that rotates eighteen times per minute and generates fifty horsepower. Powering the present machinery is an underground shaft that enters the mill basement where all the gears and pulleys are located. The mill is still operational. For more information call 764-647-6512.

Story by Jim Hamill.
Historic black and white photo by Department of Natural Resources, State of Indiana.
Color Photo by Donald J. Voelker.

Old Stump Mill
Indiana

Old Stump Mill
Near Salem, Indiana
Hohenberger Photo
Courtesy Society for the Preservation of Old Mills

Spring Mill Gristmill

Indiana

Spring Mill State Park near Mitchell, Indiana is the site of the Hamer Mill, built in 1817 by Cuthbert and Thomas Bullitt of Louisville, Kentucky. The limestone mill stood strong through constant use until the late 1800s when it was abandoned. After many years the State of Indiana restored the mill as part of Spring Mill State Park. Hamer Mill is the nucleus of a small settlement that later came to be known as the village of "Spring Mill". The gristmill stands forty feet by sixty feet by fifty feet. Three foot thick stone walls absorb the vibration of the twenty-five foot overshot wheel.

With the restoration of the mill (1927-1930), the original French buhrs began to turn again and grind meal from corn. By 1823 the Bullitt Brothers sold the gristmill and surrounding village to William & Joseph Montgomery, who added a vertical saw mill operation, distillery, a tavern to accommodate the travelers on the stage line passing through the small village, post office, general store, and apothecary to house Dr. Lemon and his practice.

As village supervisors and millers for the Montgomery Brothers, Hugh and Thomas Hamer purchased the village in 1831. Hugh became postmaster and renamed the post office "Spring Mill" and the brothers continued to operate the growing settlement through the mid 1800s.

Jonathan Turley, the last proprietor of the gristmill in 1881, installed modern milling equipment, established a commercial lime burning business and upgraded the distillery business. He maintained a stable economy until his death in 1896. This was the end of Spring Mill Village as a settlement.

Today surrounding the gristmill are nineteen other structures restored or reconstructed representing the 1832 village that flourished through the mid 1800s. Two of the residences built in the early 1800s were moved into the village from nearby because of their unique design. The two residences built by the Bullitt Brothers that served as homes of Hugh and Thomas Hamer were reproduced using other cabins. The formal garden with heirloom flowers and herbs lies near the residences. The distillery and tavern have been reconstructed. Also rebuilt is the hat factory, now used as a woodworker's shop. The post office, general store and apothecary stand as they did in the mid 1800s.

The Spring Mill Gristmill is still operational and eight interpreters bring the small village to life. For more information contact Spring Mill State Park at (812) 849-4129.

Story and photo courtesy of David L. Williams.
Photo by R.C. Holmes.

SQUIRE BOONE'S MILL

INDIANA

Squire Boone's Mill was built 1804-1809 on Buck Creek, by Daniel Boone's younger brother, Squire. The water to turn the mill wheel came from a huge spring, which flowed from a nearby cave. Other mills soon appeared in the vicinity, but used the water of Buck Creek for power.

Squire, a Baptist minister, carried his religious beliefs into the building of the mill by carving religious sayings, birds, vines, etc. into the stones of the mill foundation.

Squire Boone's Mill operated until 1852 when destroyed by fire. The McCarty family rebuilt and continued operations until the mill again burned in 1920.

The area surrounding the mill ruins and cave remained a virtual wilderness, accessible only by foot or four-wheel-drive vehicle, until the early 1970s. Fred Conway, enthralled by the beauty of the cave's interior, decided to develop it. A great deal of work, including paved walkways, steel bridges and indirect lighting was done. The cave opened in May 1973. Since then a headquarters cabin and replica log village have been added at the cave entrance.

However, the ruins of the mill foundation remained untouched until 1979, when a replica of Squire Boone's Mill was constructed. The 1850s hand hewn log cabin, originally built by Craven Boone, a descendant of the Boone family, has an eighteen foot overshot wooden wheel and a 150-foot wooden flume that carries water from the cave. The dam is built inside the cave, and the water piped through a huge pipe to the flume, with a drop of approximately thirty feet.

A set of stones, from a mill in White Cloud, grinds the cornmeal. Ted Sawle, who built and operates Hydes Mill near Ridgway, Wisconsin, built the wheel.

The Conways decided to build the new mill on the original foundation stones. The stones carved by Squire Boone (those which have not been carried away by souvenir hunters over the years), have been preserved and are on display in the second-floor museum in the mill. *Old Mill News* April 1980. For more information call (800) 234-1804 or (812) 732-4382.

Stockdale Mill

Indiana

Roann (Wabash County), Indiana
Photo by Donald J. Voelker

John Wood's Mill

Indiana

John Wood's Mill (aka T. J. Cullman Mill)
Built 1876
Hobart (Lake County), Indiana
Photo by Donald J. Voelker

Costello's Old Mill

Iowa

Built in scenic Jackson County, Iowa in 1867 by Mr. Joseph Willy, Oakland Mill, as it was then called, was later sold to Seneca Williams who operated it until 1904. The mill served Jackson and Clinton County farmers by grinding approximately fifty bushels of cornmeal and flour per day. Oakland Mill, one of twenty-six mills in use in the county was abandoned for use as a horse barn after 1904.

The fact that the building survived years of abuse stands as a testimonial to the craftsmanship of the builders. The three foot thick stone walls stand on a solid stone bluff. Inside are hand-hewn oak beams over a foot thick. The building stands three stories high.

In 1978, Patrick J. Costello, an artist, was introduced to the old, deteriorated stone gristmill. The first floor had rotted out and there were no windows, no plumbing, no wiring. The roof needed to be completely replaced. What the building did have was plenty of mice, bats and raccoons! Patrick fell in love with the old mill.

The Costellos bought the building and began the extensive renovation and the addition of several rooms and modern conveniences while maintaining the traditional lines of the old mill enabling it to be placed on the National Register of Historic Places.

The old mill is now home to Patrick, Joan and their two children, Tracy and John. It also houses an art studio.

At one time there were over 700 mills in Iowa. Now there are less than twenty and only a handful are being restored.

"Mills were an absolute necessity for Iowa's farming economy in the late nineteenth century," says Costello. "They were like gas stations are today. You just didn't get very far without one."

For more information call (319) 652-5971.

Story and photo courtesy of Patrick Costello.

Danish Windmill

Iowa

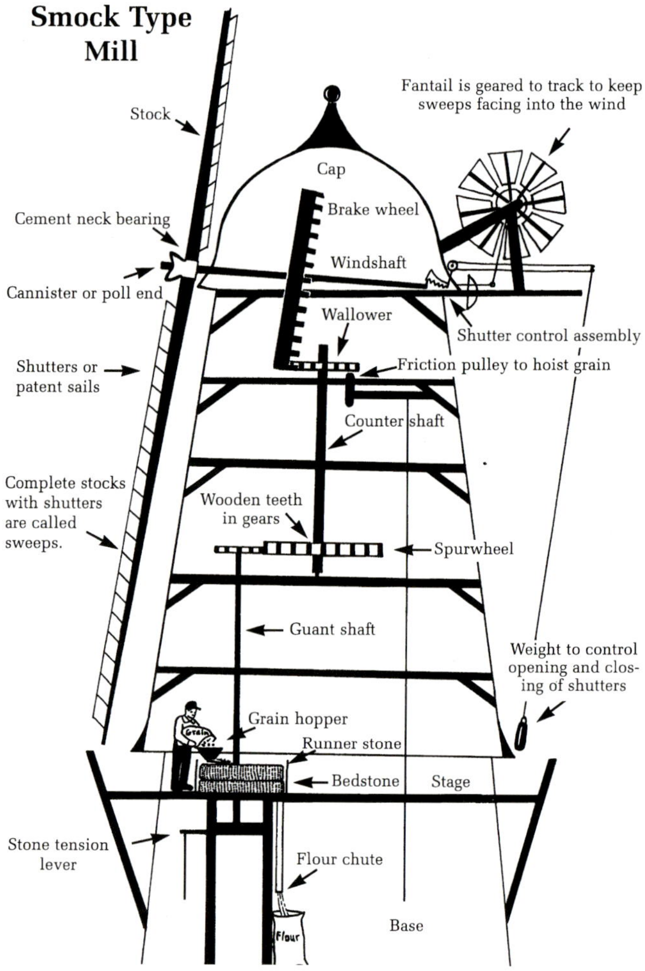

Standing tall, the Danish Windmill is a meaningful tribute to the Danish immigrants that settled here in the late 1800s. The Danes are proud of their heritage and happy to see the reminder of their homeland.

It all started when Harvey Sornson came back from a trip to Denmark in the spring of 1975 with an absolutely crazy idea. He wanted to bring a windmill to Elk Horn where the largest concentration of the Danes in America had settled. After visiting with many of the local Danes, the decision was made to go for it.

In August of 1975, the windmill project took a giant step. Harry Petersen and his wife, from Denmark, were visiting family in Elk Horn and became very interested. In September, a letter from Petersen stated that he had located a mill in Norre Snede, Jutland for sale. Estimated costs were $11,000 for the mill, $14,000 to dismantle it and $8,000 shipping by ocean freight.

Donations and pledges provided the funds sent to Denmark in early October, just five months after Sornson's "crazy idea" was born. Under Petersen's supervision, the 127 year old mill was carefully taken down. As they dismantled the mill each piece was numbered and a scale model built, six feet high, with the model parts numbered to correspond with the actual mill. The disassembled mill was loaded on a flat bed truck with the model on top. It also included all of the gears, shafts, mill stones, scales and grain carts. It left Denmark in November 1975 and arrived in New York the last week in December of that year.

After many unforeseen problems and added expense, it finally arrived in town on February 4, 1976.

It took many months of volunteer labor before the mill became operational, but what a majestic sight on a windy day! For more information, call 1-800-451-7960.

Story and photo courtesy of Danish Windmill Corporation.

Lidtke Mill

Iowa

In 1857, A.D.C. Knowlton sold to Melvin Monroe Marsh approximately eighty acres of land platted on both sides of the beautiful Upper Iowa River. The same year Mr. Marsh built the first dam and began construction of the mill. By autumn he had progressed far enough to have waterpower, a waterwheel installed and connected, and enough power to saw timber. Considerable lumber cut from the nearby woods helped fill the needs of the early settlers, who used it for hastily built shelters. Some larger timbers were used as supports in the mill. Marsh owned and managed the business successfully for eleven years. The first religious meetings in the area and those of various factions during the Civil War were held in the mill.

In 1871 the property sold to the Drake Brothers, who enlarged the operation and ground wheat night and day. Flour of superior quality was shipped to Eastern and European markets, as well as local outlets. Lime Springs became famous for buckwheat flour. An infestation of chinch bugs spelled doom for miller and farmer alike, followed by a disastrous crop failure in 1878.

However, the Depression led to a succession of owner-operators at the mill, D.W. Haven in the early seventies with R. Baldwin as miller, H. W. Babcock in 1879, joined by A. F. Baumgartner, F. M. Clark and Dr. John W. Reed, then Dr. Reed and his son-in-law, George Heath. Once again the old mill became profitable, with six full-time employees. Among these were Heath, engineer Gid Quinn, one of the Hewes boys, Dirk J. Kleiboer and Fred Salisbury who was twice millwright and manager in early 1900 and again 1909-1913. Roy Nash worked there as did "Big John" Davis, Charles Schacht and Benhardt Roesler. Fire brought an abrupt end to this prosperity in 1894. The same year James Safford bought the site and built a mill on the original foundation. Now, feed grinding and generating electricity were the specialties.

In 1905 William Freuchte became owner and in 1915 it again changed hands when Dave W. Davis bought out Mr. Freuchte. Mr. Davis enlarged the mill, built a new dam and placed a brick veneer on the mill exterior. Herman Bundrick served as miller until 1919. The mill provided electricity in the evenings for Lime Springs and Chester and, on certain mornings, for washing and ironing. In time, Herman Lidtke, son-in-law of Davis, became owner and continued the mill operation until 1960. Other employees were Enos Wilken, Bill Mohs, Willis Lloyd, Emerson Jones and Willard Lloyd. Lidke Mill Historic Site is now a ten acre park across the Upper Iowa River and Dam. For more information call (319) 566-2827.

Story and photo courtesy Valery Gebel.

Motor Mill
Iowa

The dream of Motor Mill had its beginning in the early 1860s. John Thompson, J.P. Dickinson and James 0. Crosby formed a partnership to build a gristmill, sawmill, farm and town site at Hastings Bottom, the site of an earlier sawmill. The men changed the name of the settlement to Motor, for reasons unknown.

Thompson and his associates spent $50,000 to build Motor Mill, and $40,000 on equipment and outbuildings. The mill and cooperage are believed to have been built between 1867 and 1869, the other buildings, in the early 1870s. The scenic iron bridge over the Turkey River was constructed around 1895.

German masons from the nearby village of Communia were hired to build the mill. Stones for the buildings were quarried on the top of the bluff, and lowered on small cable cars to the building site. Legend tells that four "lead" masons were responsible for one wall of the building, and that each tried to outdo the other with his own style of stonework. This may explain why the stones on three sides of the mill have rounded faces and the fourth side, square faces.

A dam upstream, about 200 feet long and twelve feet high, provided water for mill operation. A fifty foot flume directed water into the basement of the mill, providing power for four turbines.

The large dam and turbines insured a minimum of 250 horsepower even at low water. One of the owners boasted the mill could handle 1500 bushels of wheat per week.

According to Clayton County history, during construction of the mill, laborers were going to hang Thompson for failing to pay their weekly wages. Another early yarn indicates the men were so angered they threatened to blow up the dam if the payroll was not met.

The town of Motor was quite prosperous for a time. The building of a railroad began, which would link Motor with the town of McGregor. But an 1875 flood washed out the ties and rails, destroying any hopes of rebuilding the spur.

The mill suffered another blow when large numbers of chinch bugs and other small insects inflicted great damage to local corn, grains and grasses. Invasions of insects in 1867, 1871 and 1887 destroyed virtually every acre of wheat in northeast Iowa. Motor Mill ceased operation in the 1880s and sold for a mere $12,000.

Many changes occurred after the closing of the mill. These included gutting the mill, selling milling equipment for junk, converting the stable into a dairy barn in the 1930s and raising the roof to make more room for hay storage. The buildings were used in dairy operations until 1982.

The majestic old mill is approximately ninety feet tall, and believed to be the tallest of its kind in the Midwest. The stately structure boasts six full stories. The foundation walls of the building are five feet thick, gradually decreasing toward the roof.

Photo and story courtesy of Chuck Carbert.

Pine Creek Gristmill

Iowa

On September 21, 1832, a treaty signed between the United States Government and the Sauk and Fox Indians, occupying eastern Iowa, became known as the Blackhawk Purchase. According to terms of the treaty, the Indians were to give up possession of the land on June 1, 1833, thereby opening up land west of the Mississippi River for settlement by the white man.

In the spring of 1834, the first permanent white settler, Benjamin Nye, came to what is now Muscatine County, Iowa. He built a cabin at the mouth of Pine Creek near the Mississippi River and opened a store with goods he purchased in St. Louis. This was a tumultuous period in Iowa history with the surge of settlement into eastern Iowa. The country rapidly transformed from wilderness into farms and villages. To supply the influx of settlers, Ben Nye built a sawmill on Pine Creek approximately one-half mile up from the Mississippi River. The exact date of construction is not known, but is thought to be 1835. By 1837 the sawmill had been moved upstream another one-half mile and a gristmill built across from the sawmill.

The steady inflow of settlers soon overwhelmed this early gristmill, and in 1848 Ben Nye built a larger gristmill. This mill, Pine Creek Gristmill, is now in Wildcat Den State Park

There is no trace left of the earlier mills, but the Pine Creek Mill and the milling machinery inside have been preserved. In construction of the mill, Nye used native lumber cut in his sawmill. At the time of construction the timbers used were the largest in any building in the territory, some fourteen inches square. A post and beam construction technique was used, with native oak flooring and weatherboards of walnut. The overshot waterwheel in the basement, which originally powered the mill, had been removed by subsequent owners and replaced with a water turbine. Sometime in the 1800s a steam engine was also installed to power the mill when there was insufficient water.

Benjamin Nye did not operate his mill for many years before his untimely death. George McCoy worked for Nye as a farm hand and eloped with his daughter. Nye never forgave nor spoke to McCoy again. McCoy and Nye's daughter were living in Cedar County where McCoy became sheriff in 1840. In 1848 McCoy left the area to seek his fortune in the California gold fields. Upon returning two years later he found his wife with a new baby. During a confrontation between McCoy and Nye, involving guns, clubs and knives, Nye was killed by his son-in-law. It was ruled, however, that McCoy had acted in self-defense. Ownership of the mill then passed first to another son-in-law, and others, until the Missel brothers bought the mill in the late 1800s. The State of Iowa acquired possession in 1927.

Through the years the Pine Creek Gristmill fell into disrepair. In 1996 the Friends of Pine Creek Gristmill organized and worked with the State of Iowa to restore the mill to operating condition. The organization also staffs the mill. For more information call (319) 263-4337.

Potter's Mill
Iowa

Potter's Mill is Iowa's oldest gristmill, built in 1843 by Capt. Eldridge Geny Potter. Potter was a stonemason by trade, a Captain in the War of 1812 and a successful distiller and miller from New Lebanon, Illinois. He formed a partnership with local millwright, John Gammel, to build his "Jasper Mill" on the Big Mill Creek near Bellevue, Iowa. Using limestone from the bluffs across the creek for the foundation of the building and the eleven foot thick dam, cut timber and an additional 73,000 board feet of logs were purchased to complete the structure. Throughout the building are the original hand-hewn walnut and oak beams, forty-five feet long and fifteen inches square. An overshot waterwheel, twenty feet in diameter with twelve foot wide wooden buckets, powered three huge four-foot wide "runs" of buhrstones and all the mill equipment. In the 1860s, installation of six turbines in the sub-basement required that the limestone dam be raised to its current height. Water poured through the four foot high archway in the dam and fell fourteen feet to the turbines, which in turn ran six sets of milling stones. Peak flour production at the time was 200 barrels per day.

E.G. Potter sold the mill in 1871 to Kilborn and Company. The Kilborns and subsequent owners modified the mill to keep pace with changing technology. The 1880s saw the introduction of twelve roller mills. On May 24,1896, a flash flood, caused by fourteen inches of rainfall in twelve hours, rushed down Big Mill Creek knocking out a dam upstream and tearing out the mill spillway and a frame office. Instead of rebuilding the spillway, a twenty-five horsepower steam engine supplied power until the use of electricity in the early 1900s. The mill never regained its former capacity. Able to power only two runs of stones, production capability fell to fifty barrels of flour per day.

Potter's Mill remained active until 1969 when the machinery ground to a halt. The old mill sat beside the stream and Mother Nature started her deterioration process over the next twelve years. In 1980, Daryll and Carolyn Eggers of Marshalltown, Iowa, bought the building at auction. Reconstruction began on the creek side foundation that had crumbled dangerously. The Eggers made an extraordinary and successful effort to restore the mill to its original form. In the 1980s, Potter's Mill received designation on the National Register of Historic Places.

In November of 1998, Larry and Kathryn Howell, enterprising teachers and coaches from the lower Rio Grande Valley of Texas, purchased the mill. The entire building is utilized to its fullest. It now contains a dining area, pub, antique mall and bed and breakfast. For more information call (877) 295-3412.

Story and photo courtesy "Coach" Howell.
Photo by Sandy L. Dyas.

Rhodes Mill
Iowa

Fertile, Iowa
Built in 1878 by William Rhodes.
Photo courtesy Steven L. Rasmussen

Wagaman Mill
Iowa

The Wagaman Mill in Lynnville, is one of few remaining mills in Iowa still able to operate with waterpower. Established along the North Skunk River by John Sparks, the first mill was a small, paddle wheel sawmill with a wooden dam. Construction began 1845 and was completed in 1846. In 1847, John Sparks began building a gristmill, completed in 1848. The buhr (grinding) stones of Spark's Mill were said to have been of the best French make. The mill sold in 1851, and passed through many owners until 1898, when the Wagaman family purchased it.

In 1868 Jesse Arnold, then owner, ordered a vertical drive hydraulic turbine. This installation revolutionized the Iowa milling operation, eliminating the old large paddle wheel, and allowing continuous operation twelve months of the year. Prior to the turbine, the mill had to close due to winter freezes. A steam engine was installed in the mill around 1900 and replacement of the wooden dam with concrete and a second waterpowered turbine were added in 1919.

The last owners to use waterpower were the Wagamans, who operated the mill for three generations and the name Wagaman Mill has been permanently attached. Installation of modern flouring machinery occurred during this time. The mill reached peak production during World War I.

Commercial flour became readily available in the 1920s and operations changed to a combination feed mill and electric power production. The "Old Mill" supplied between 5,000 to 15,000 kilowatt-hours of electricity each month to the surrounding population. In 1958 use of waterpower ceased and all electric machinery was installed. The old turbines were left in their pits.

The mill has operated as a sawmill, wool carding mill, flouring mill, feed mill, and power generating sub-station. It has also been entered on the National Register of Historic Places.

Jasper County Conservation Department took responsibility for restoration of the mill and waterpowered equipment. For more information call The Jasper County Conservation Department at (515) 792-9780.

Story and photo courtesy of Mark D. Wagner.

Champeny's Old Oxford Mill
Kansas

Champeny's Old Oxford Mill is a three-story stone building forty feet long and thirty feet wide, with a basement. The steep gable roof is surmounted by a wood frame monitor called a "Texas", added when the mill converted from the stone ground method of making flour to the roller mill process. Later additions include a small packing room and warehouse attached to the north side.

Shortly after 1870, when the Osage Indians in Kansas moved south to Indian Territory, the town of Oxford incorporated. By 1873 Mennonites brought turkey wheat from Russia and began growing it in this area. The locals needed a flour mill so they would not have to haul wheat great distances. Judge John M. Hewitt from Red Oak, Iowa, provided the financing, and the Sumner County surveyor, D. N. Cook, selected the site for the mill. A citizens group took responsibility for construction of the millrace at an estimated cost of $2,500.00, primarily raised by subscription. Mr. Sleigh, donated the stone for the $15,000 mill and the 1000-foot long dam across the Arkansas River to allow water to be diverted into the millrace.

The mill was built in 1874, from a design that included arched openings on two sides to allow water to flow through the mill as a power source. These openings, however, were never used and no wooden waterwheel was used for the mill. Instead, two Leffel turbines were installed in the millrace just outside the building. Power generated by the turbines drove the stones that ground the wheat into flour. The mill design could accommodate four runs of buhrs, but milling hard winter wheat on stone buhrs did not produce satisfactory flour, as with the earlier soft wheat. The design and competition from mills in Cowley County forced the installation of the new roller mill process in 1885. A 350-kilowatt generator was added in 1935 and the waterpower used to generate electricity for the newer mill built nearby.

The Kansas State Board of Agriculture in its 1875 Annual Report listed 290 flour mills in Kansas. Of this number, one half were steam powered, ten were wind driven and the rest, like the Old Oxford Mill, waterpowered.

In 1900 Kansas had over 400 small mills. By 1948 there were only eighty-six mills but they produced a greater volume of flour than the greater number of mills produced fifty years earlier.

Charles Champeny and B. F. Fisher purchased the mill in 1910. In 1916, Fisher sold his share to the Champenys who have maintained their interest in the mill. However, in 1935, a mill closed in Claflin, Kansas and was moved and rebuilt near the old mill. The old mill then became neglected and in a state of disrepair.

Champeny's Old Oxford Mill gained designation on the National Register of Historic Places, April 26, 1981. Wallace Champeny and Hal Ross, graduates of the flour milling school at Kansas State University, Manhattan, Kansas, started restoration in July 1988. The National Park Services certified that the project met the "Standards of Rehabilitation" on October 28, 1991.

The flour mills remaining in Kansas can be counted on the fingers of both hands, yet Kansas produces more flour than any other state in the nation. Today the mill houses a restaurant, lounge, gift shop and small museum. For more information call (316) 264-4981.

Story and photo courtesy Hal Ross.

The Friesen Dutch Mill

Kansas

The newest addition to the historical complex in Hillsboro, Kansas is the Friesen Dutch Mill. This replica of the Jacob Friesen Mill, originally built in the Russian Mennonite village of Gnadenau in December of 1876, is fulfillment of a longtime dream of A. E. Janzen, Solomon L. Loewen and other Hillsboro historians. The wind-driven Dutch gristmill ("hollaendische Muehle"} represented a continuity with the ethnic heritage of the early German-speaking settlers of Marion County. It is a symbol of their migration from the Netherlands to Prussia to Poland to Russia to America. Jacob Friesen, an experienced miller, built the original mill. He immigrated to Kansas in 1874 from Ohrloff, Russia. The local newspaper reported soon after the construction of the mill, "The gristmill at Gnadenau is running day and night when there is wind. They grind corn, rye, barley and wheat...(March 1877)... The Gnadenau gristmill has all it can do (February 1878)". This Dutch-style mill was the second one to be built in the state of Kansas, second only to the mill at Lawrence, Kansas (1862). Later Dutch mills were constructed at Schoental, Wamego, Smith Center and Pfeifer, Kansas by Dutch and German immigrants.

The 1974 Kansas Wheat Centennial provided some impetus for this Hillsboro dream, but not until 1990 did the dream began to take on substance. At that time Dr. G. George Ens's drawing of the Friesen Mill was selected as the winner of the Emprise Bank calendar contest, resulting in the distribution of hundreds of Friesen Mill calendars. Vernon Wiebe, chair of the Hillsboro museum facilities development committee, enthusiastically promoted the mill project, and Marvin Jost contacted members of the Jacob "Miller" Friesen family regarding the construction of this replica.

Richard Wall accepted the challenge of designing blueprints for this mill replica in the spring of 1992, based on a single photograph of the original mill. Construction began in April of 1992. Richard Wall and volunteers worked throughout the summer to complete the post and beam framework by winter. Two years later, on May 29, 1994, the Friesen Mill was dedicated as part of the 110th anniversary celebration of the city of Hillsboro. Descendants of the Jacob Friesen family came from all over the United States to attend this celebration. At this time Peggy Goertzen also released a 182-page history of the Friesen Mill and early milling in Marion County, Kansas.

Richard Wall and Paul Friesen installed new millstones on the second floor of the mill the following summer. Since that time, the mill has intermittently ground whole wheat flour on occasions of sufficient wind.

The Friesen Mill is part of the Hillsboro Heritage Park and the property of the city. It is managed and operated by the Hillsboro Historical Society and Museum. For more information call (316) 947-3775.

Story courtesy of Peggy Goertzen.
Photo by Orley Friesen.

Smoky Valley Roller Mill

Kansas

The Smoky Valley Roller Mill, built in 1898 along the Smoky Hill River in Lindsborg, is the third built on this site. C.J. Johnson built the first in 1872. The simple grist and sawmill washed away in a flood in 1880. New partners J.G. Bergsten and Francis Johnson opened a second mill in 1881. The facility became a roller mill in 1889.

Theodore Teichgraber purchased the mill in 1890. Teichgraber came from a milling family and had brothers in the business across the state. Known as “the jolly miller”, Teichgraber ran a successful business for a number years.

Disaster struck in 1897 when a fire destroyed the mill. The company rebuilt and the current mill opened for business in April 1898. Two water turbines, generating 125 horsepower initially powered the mill. The mill processed thirty to thirty-five bushels of wheat per hour, producing between 1260 to 1470 pounds of flour per hour. Four to six workers were required to operate the mill. At least five different brands of flour were produced.

Unfortunately, the miller’s untimely death in 1907 brought changes. William and Emanuel Hagstrom purchased the mill and operated it until 1927, not always as successfully as Teichgraber. In 1927, the mill sold to Art and Clarence Runbeck. In the 1930s the Runbecks converted to electricity for power. By diversifying the product line, the business remained operational through the 1940s. By the 1950s many improvements were needed for the mill to survive with new and larger competitors. With the death of the last Runbeck brother, the family opted to close in 1955.

In 1962 the property was given to McPherson County for preservation. Volunteers began cleaning and planning the mill restoration. In 1972 the Smoky Valley Historical Association took over the effort and the mill was listed on the National Register of Historic Places.

In 1974, George Tesarek, a retired miller, from St. Joseph, Missouri, became technical advisor. Tesarek’s milling expertise and knowledge of the industry proved invaluable. After seven years of hard work the machinery had been fully restored. In 1981, the mill became operational once again.

Today, the Smoky Valley Roller Mills are part of the McPherson County Old Mill Museum. The mill operates once a year and the sights and sounds of the mill in motion never fail to leave a lasting impression. For more information call the museum at (785) 227-3595.

Story and exterior photo courtesy of Lorna Patterson.
Exterior photo by Tom Jaderborg.
Interior photo of mill equipment by Michael Snell.

Wamego Dutch Mill

Kansas

The century-old Dutch Mill, a familiar Kansas landmark, dates back to 1879, when a young Holland immigrant observed farmers in Kansas still grinding wheat and corn by hand. With this in mind, John Schonoff built a sturdy windmill on his farm twelve miles north of Wamego. Wooden mechanisms were designed and built in his workshop and the mechanical iron parts were hammered out on a homemade forge. Horse drawn wagons brought the main drive shaft, the only part not made by Schonoff, to the farm from Leavenworth.

John Chadwick, a stonemason from Wamego, erected the tower, from yellow limestone quarried on a nearby farm, using white limestone around doors and windows as trim. The building rose forty feet, with a diameter of twenty-five tapering to twenty feet above. The statue of Ceres, Goddess of Grain, adorns the wall of the second story window. Done by Chadwick in secrecy, he would hide it in the grass if approached. It is believed that the features of the Goddess were modeled after Schonoff's daughter.

Ed Regnier purchased the farm in 1896 or 1897, and used the mill for storage but upkeep became a problem. When A. M. Bittman and Forest Leach proposed the donation of the mill to the city park, Regnier approved. On June 6, 1924 the mill, disassembled stone by stone and numbered, was ready for the move. Thirty-five horse drawn wagons hauled the mill into town for reconstruction in the city park.

Sixty-four years the mill has stood on the twenty-five foot hill beside the lake in Wamego City Park. Thanks to John Landes, the Wamego Historical Society, the City of Wamego, the Wamego Tourism Committee, and donations from private citizens the mill reopened July 4, 1988 for grinding. For more information call (785) 457-7849.

Photo and story courtesy of Wamego Area Chamber of Commerce/Main Street.

GUYN GRISTMILL
KENTUCKY

The Guyn Gristmill, built in the 1870s, is a two and a half story steampowered gristmill. Forty feet by thirty feet, it has massive tennoned beams and frame, circular-sawn rafters and joists. The wall treatment is twentieth century clapboard and the foundation is ashlar cut Kentucky limestone with stone piers. Two large buhrstones are situated on the first floor. Although deteriorated, many sifters, blowers and augers are still in place. Many of the wooden gears and a cogwheel are in the basement. The steam engine is attached to a cast iron boiler and a limestone wall, previously supported by a protective structure, surrounds both.

The waterpowered Guyn Sawmill, built 1840-1850s, is a rectangular plane, with a braced frame fifteen feet by forty-five feet. The raised limestone foundation is dry laid with several original bays and a crude rubble limestone ramp. The mortise and tennon acize beam framed structure has three heavy bents and pole rafters. Logs entered in the west and exited at the east gable.

The Guyn Historical District complex ceased operation in 1920. Mrs. Clay Brandenburg owns the Guyn Historical District. She is the seventh generation in the William Rankin Guyn Sr. bloodline to own the land. This area is listed on the National Register of Historical Places. The gristmill is currently a studio and woodshop for Prajna Design/Construction, who have carefully mended the mill over the past fourteen years.

Story and photo courtesy of Mary Putnam.

Hurts Mill
Kentucky

Mt.Piscah (Wayne County), KY
Photo by J.C. Barger

Irvington Mill
Kentucky

Irvington (Breckinridge County), KY
Photo by J.C. Barger

McHargue's Mill

Kentucky

William McHargue I and his family left North Carolina in 1805 and traveled into Kentucky over the old Wilderness Road, settling in what is now the southern part of Laurel County. On the bank of Lynn Camp Creek, he erected one of the first water-driven corn and sawmills in southeastern Kentucky. William McHargue II built another mill between 1817 and 1820 on Robinson Creek where the Wilderness Road crossed it, and members of his family operated the mill until the 1880's.

In 1937, the museum area of the Levi Jackson State Park began the early stages of development around a "mountain life" or pioneer theme. The local park committee, which at that time included Russell Dyche and Lee B. McHargue, proposed construction of additional buildings in that area.

The mill was built in 1938-1939 on the banks of the Little Laurel River where the old Boone Trace crossed it. Before drawing plans, the architects had visited a number of old watermills in this area to insure that it would be true to type. In essential interior parts, including the millstones, the mill dates to the original McHargue family mills of the early 1800s.

"McHargue Mill" operated sporadically until the early 1960s. Restoration in 1989-1990, as a special project of the Kentucky Department of Parks, has rendered the mill operational once again. For more information call (606) 878-8000.

Story and photos courtesy Carol Kirby.

McHargue's Mill
London (Laurel County), KY
Photo by J.C. Barger

Rogers Mill

Kentucky

In a hidden valley of unusual beauty in Summer Shade, Kentucky, Rogers Mill is built on the foundation stones of prior mills. It is named for Edmund Rogers who once owned the land on which the mill stands. As a young lad Rogers served as messenger in General Washington's headquarters at Yorktown. He later became chief surveyor of the Kentucky lands reserved south of Green River for Virginia's Revolutionary War veterans.

At least sixteen millers are known to have operated mills at this location over the years. The mill took the name of the miller, but has also been known as the Cave Mill. Frequent changes in ownership would serve as good indication that the mill was never very profitable. The miller probably had additional income from farming or other interests. At times there were competing mills; a horse mill at the head of Fallen Timber Creek, and a small mill on the creek for grinding corn, which washed away and was not rebuilt.

The information gathered about the last operating mill comes from personal memories of those who lived in the area at the time. There is little recall of the mill interior or grinding process, but the rugged terrain, the dark cave, and the towering flume that carried water to the huge creaking, splashing waterwheel made a lasting impression.

Milled, or cracked corn deteriorates fairly rapidly. This made it necessary to visit the mill on a regular basis. A neighbor, who still lives nearby, recalls the occasion his father first entrusted him to the sure-footed mule that carried him, astride a half-filled bag of shelled corn, safely down the steep hillside to the mill. The miller took the corn and from the meal on hand, measured out the amount due the boy after extracting the proper toll. The toll system allowed a standard amount of meal, usually one-eighth, to be retained by the miller for his services, a convenient method of payment in a time when money was scarce. It also served the miller well. Millers have traditionally kept fine livestock. The animals were fed the meal collected in excess of the miller's personal requirements. The miller is said to have fattened hogs, which he kept in the large rock-shelter on the hillside above the cave.

"Going to mill" was usually considered a pleasant chore. Often a number of customers arrived at the same time, which may have been intentional, for they enjoyed the relaxation and conversation with neighbors. The miller waited upon his customers in order of their arrival; and the processing of their grain into meal was called a "turn", that is, a turn of the grinding stones. From this comes the still common expression, "Now it is my turn."

Just keeping the previous mill in operation during its final years became a constant battle, because the number of customers declined and repairs mounted. Around 1930 a new waterwheel of white oak was built. The pattern for the wheel was still in existence a few years ago, but the wheel, never installed, was left to rot away on the ground. The mill no longer operated in 1935, and former customers carried their corn several miles away to be ground on a gasoline mill. This became the fate of thousands of watermills across rural America.

Rogers Mill now serves as a museum telling the story of corn and the part played by the community mills before the automobile and modern packaging. For more information call (270) 428-2592.

Story and photos courtesy of S.R. Gorin.

Rogers Mill
Summer Shade (Metcalf County) KY
Photo by J.C. Barger

Mill Springs Mill
Kentucky

Monticello (Wayne County), KY
Photo by J.C. Barger

G.M. Taylor Mill

Kentucky

Built in 1889, as the Bluegrass Cordage Co., at the staggering cost of $100,000, the design allowed the mill to convert two million pounds of hemp annually for the twine industry. The employees and equipment were originally from Scotland. The factory, self-contained, had an electrical plant, fire department and railroad spur to the main line. It is listed as one of the largest factories of this type in America.

In 1923 W.J. Askew, father of Mrs. Emily Rawdon, purchased the facility. She refitted and converted the mill into a farm supply operation. At a time when coal was king, Georgetown had a total of ten coal dealers. G.M. Taylor Seed Co. is the only survivor of this once very competitive business. Wool was also bought and sold in large quantities.

Born in 1859, the legendary G.M. Taylor made a name for himself stripping Bluegrass Seed. The company, originally located on W. Main, in later years moved to Maddox Street.

In 1941, W.J. Askew and Taylor formed a partnership and Taylor moved his business to the present location.

Seed cleaning became the primary business and had customers all over the U.S. and parts of Europe. After sitting idle for years the original antique cleaning equipment has been rebuilt and is in full use today, thanks to miller Levi.

Built over a century ago, the mill survived changing times, devastating fire and the loss of the original founders.

The main building, 42 feet by 178 feet, is made of seven course common bond brick in the Victorian Industrial Style. The safe was built in Cincinnati in 1888, when few trusted banks. Each door weighs 4000 lbs. for a total weight of 19,500 lbs. The mill is still family owned and operated.

Story and photo courtesy Gaylord and Tracy Hilander.

Weisenberger Mill

Kentucky

Weisenberger Mill site, situated on South Elkhorn Creek outside the small town of Midway, is more than healthy trees, sparkling water and abundant wildlife. Weisenberger Mill has been an institution in the milling business, and the waters of South Elkhorn Creek have powered the mill's turbines for over a century.

There has been a mill on this site since the 1800s. "We've been here for a long time" says Philip Weisenberger, whose great-grandfather, August Weisenberger, emigrated from Baden, Germany in 1862. "It's a family business and we're right proud of it".

A concrete structure replaced the original mill in 1913 and additions were made through the years, the latest in 1998. In the early years the mill primarily produced soft wheat flour and white cornmeal. In the 1950s "add only water" mixes were introduced.

Six generations of Weisenbergers have proudly carried on the family tradition of milling wheat and corn on the Elkhorn. This is the oldest commercial mill in Kentucky operated by the same family without interruption.

The mill site was placed on the National Register of Historic Places in 1976. For more information call (606) 254-5282.

Story by Lori A. Smith.

Wolf Pen Branch Mill

Kentucky

Prospect (Jefferson County), KY
Photo by J.C. Barger

LEONARD'S MILLS

MAINE

Leonard's Mills, a reconstructed mill settlement, is the centerpiece for the Maine Forest and Logging Museum's efforts to preserve and celebrate the forest and settlement history of Maine. Located a mile upstream from the mighty Penobscot, the Museum has rebuilt a stone dam on Blackman Stream (original c.1797) and reconstructed an up-and-down sawmill. The water flows through a gate and canal into the millpond to power the overshot waterwheel. It is a rare working example of a technology common throughout the colonies, from the 1600s to about 1840 (and in limited use well after this). Mills such as this could cut between 500 and 3,000 board feet of lumber per day, providing essential building materials for home, farm, and industry, as well as rough-sawn boards, timbers, clapboards, staves, and deal for export.

The waterwheel is made of white oak, hemlock, cedar, and spruce. Its diameter of nine feet allows for the development of six horsepower. The gearing includes a "bull wheel," constructed of fifty-eight rock maple teeth, which drives a sixteen spoke "trundle wheel." These wheels drive the "eccentric wheel," which is attached to the "crank arm." This is attached to a "pitman arm," which drives the "saw sash" up and down thirty-six strokes per minute. Ideally, at this rate, it takes about sixteen minutes to cut each board from a twelve foot log.

Featured in the settlement is a covered bridge, an 1850s era blacksmith shop, a reconstructed log cabin, a trappers line camp, a hovel, and more. For more information call (207) 581-2871.

Olde Gristmill
Maine

In 1771, the original proprietors of what is now Lebanon, saw the need for a mill site in the eastern section of the township. They voted 200 acres of land be sold for about $500. It is possible that the mill was built as early as 1774 by Joseph Hardison, who also established an up-and-down sawmill further down river.

In 1785 a spring freshet took out the mills and bridges along the Salmon Falls River, but the mill and the bridge at Little River Falls survived and ground grain for the inhabitants of the town until the other mills were rebuilt. Both the mill and bridge are on the National Register of Historic Places. The mill is believed to be one of the oldest wooden mills in the country. It was also the site of the first gas pump in town.

The gristmill has changed very little through the years. It was framed with timbers measuring twelve feet by twelve feet and thirteen feet by thirteen feet, hand-hewn and cut on an up-and-down saw, probably in the mill down river. The hopper to hold the grain had to be enlarged when the mill started receiving grain from the midwest. Grain would be shipped to Lebanon by railroad and hauled by wagon two miles to the gristmill. A store added to the back of the mill extended out over the river. The well grooved steps lead to an entrance through the unique old folding door and a journey into the past.

The gristmill could produce 200 barrels of meal a day, evident by the well-worn floor where the miller stood during the operation. An undershot waterwheel below the mill once provided the power for the two runs of stone, a buhr and a local granite. The turbines that replaced the wheel are still there. The wooden gears, pulleys, shafting and elevator cups are no longer in operation.

An extension of the shingle and clapboard mill was also added. The production rate of the shingle machine can be seen on a board hanging on the wall, showing the low day twenty-four square and the high day 129 square. This is about 27,000 and 129,000 respectively. The figure represents two twelve hour shifts, not uncommon at the mill.

The mill is privately owned by Ted & Betty Frederick, who have renovated the shingle and clapboard mill into living quarters where they reside year-round. The gristmill remains unchanged. For further information call (207) 457-9220.

Story and photo courtesy of Betty & Ted Frederick.

Raye's Mustard Mill

Maine

Around the turn of the century John Wesley (Wes) Raye returned home from the Spanish American War. He took a job with one of the sardine companies that were the mainstay of coastal employment at the time. Sardine packers were looking for sauces to pack herring in and Wes became interested in mustard. His research indicated that cold ground mustard enhanced the flavor of food.

The company he worked for was not receptive to his ideas for producing mustard sauces and he set up his own mustard mill in a smokehouse his family owned. The mustard he produced proved to be very popular and soon a larger building became necessary. The present mill site was chosen because of the close proximity of the train tracks. The mill had a siding for unloading supplies, such as seed, vinegar, and spices. Full barrels of mustard were then loaded onto the train, transported to the steamship wharf and picked up by the sardine carriers.

Around 1938, his son, H. Wadsworth (Wad) Raye, returned to Eastport to work with him. During WWII they grew mustard seed locally fearing the war would prevent them from getting seed from Northern Europe. Wad became a partner and took over the business when his father died. The sardine industry continued to flourish and in the 1950s Wad set up another mill in Brunswick to handle sales from Rockland south. Although both father and son ground special mustards for family and friends, only mustard sauces for sardines were produced commercially before 1991.

Over the years, local people would bring their jugs to the mill to be filled with "factory" mustard, not as thick as the specialty mustards but very tasty. Many old-timers have said when they were growing up their folks worked in the factories and brought home mustard for "mustard sandwiches" during the lean times.

By 1990, increased production costs and a diminished sardine market no longer justified the business and closure became imminent. Nancy Raye, Wad's daughter in California, knew this to be the only remaining part of the once flourishing sardine industry in Eastport and probably the last mill of this type to exist. Beginning with a couple of recipes for specialty mustards, she ventured into the specialty food market.

The entire mill is run by an overhead drive shaft. An electric motor with a large belt to the flywheel drives the shaft. Prior to 1946, a coal fired steam engine served as the power source. Although the mustard passes through four stones in the grinding process, there are five stones on line. If one stone needs to be "dressed", it can be worked around. Because of the age of the mill, replacement parts are not readily available and usually have to be made when ordered. Several machinists in the area use ingenuity to keep the mill running so the mill can continue grinding into the next century. Raye's Old Stone Mill is still grinding mustard as a working museum. For more information call (800) 853-1903.

Story and photo courtesy Nancy Raye.

Smith Mill
Maine

Smith Mill in Bridgton, Maine was named after the first operator. He made doors, windows and coffins and called the mill "The Old Coffin Shop".

Construction dates to approximately 1867, as a wood working cabinet shop with waterpower. It is the only mill still standing of twelve that were built on this stream. Various people operated the mill until 1947 before leaving it to collapse.

The present owner bought the mill in 1982, jacked it up and pushed it away from the river. The shop area was used for prefabricating parts for houses and for cabinet making.

For a while the second floor opened as a wood products retail shop. Some of the early machines are still on the first floor. The third floor, converted to an apartment, is where the owner, now retired, lives.

Story and photo courtesy W. Meade Werner.

HYATTSTOWN MILL

MARYLAND

The mill at Hyattstown was built in 1798. At the time of the town's founding, the miller was George Wolfe, Sr., who had taken over the mill when William Richards died. Mr. Wolfe purchased the mill site in 1807 and sold the property to Benjamin Waters and Frederick Baker for $5,000. In 1814, Baker operated the mill until 1827 when he sold to Otho Norris and wife, Sarah. They sold the property to William Farrer, who made substantial improvements to the mill over the next five years. It was a two-run flour mill and sawmill, with one employee and an output of 6,000 bushels of meal and 35,000 feet of lumber annually, valued at $4,125.00.

In 1858 William Farrer advertised the property for sell. The published advertisement in the Montgomery Sentinel on October 8, 1858 lists *a re-built mill with new buhrs, and a group of connected buildings, twenty-five acres of land, with a large, frame dwelling house, one small dwelling house, one flouring mill, two pair of four feet buhr, new, and the latest improvements for a merchant mill. One country mill, one pair of three feet seven inch buhrs all new machinery, one sawmill connected with the merchant and country mills.*

The term "merchant" or "flouring" mill refers to two runs of stone usually reserved for grinding of wheat. While the country mill would be ordinarily used for custom grinding or grinding corn, apparently all housed in one building.

George A. Darby purchased the mill site in 1872, for $2,901.00. The 1880 census lists Darby as the owner, with one assistant. He operated the mill until 1905, then sold to John W. Harris. Harris made improvements by installing a new roller mill, adding a steam boiler and engine and an inside and outside corn shelter beneath a shed roof. Charles Basil Murphy served as miller when Harris owned the mill. He sold to Bradley Hill Dudrow in 1991, who purchased it as an investment. He sold in 1912, to Mortimer and Luhn. Ruth Mortimer Price recalls playing in the bran piles beside the mill. In 1914 Charles Russell Murphy began grinding for Frank Mortimer and Charles Luhn. Mortimer had experience as a machinist (mechanical engineer) with Baldwin Locomotive Co, Philadelphia. Luhn was a carriage maker, cabinet maker and wheelwright. Both did their work while financing the milling operation making the mill a small-scale industrial park. The mill handled grinding of some grains and cereals. Winter wheat and yellow and white corn were the principal grains milled. Most of the grain was purchased outright. Some custom grinding continued. In toll grinding the miller retained 25% of the finished product as a grinding fee. The whole wheat flour, baker's flour, and corn meal were marketed locally in neighborhood stores, transported from the mill in a truck. The mill also marketed self-rising flour, produced by milling baking powder with wheat flour. A small amount of marketing was done in the Frederick County area.

Montgomery Sentinel August 29, 1918 stated "fire from overheated steam boiler consumed the building igniting corn cobs." At the time of the fire, the mill had 1,000 barrels of wheat inside. Although a possibility, spontaneous combustion was not proven. Mortimer and Luhn had little capital. The new mill was to be built the least expensive method possible. Web Burdette supervised the construction. The community pitched in to help rebuild the mill.

The metal waterwheel was shipped to Hanover, PA for repairs. Mortimer and Luhn purchased the warehouse of Price's Distillery, and dismantled and incorporated it in the new mill. A run of buhrstones and millworks were taken from a mill in Browningsville. During the 1920s larger mills located near the railroad totally overpowered the Hyattstown Mill. The larger mills had the capacity to produce bleached flour, the most popular type, which the smaller mills didn't handle. Mortimer and Luhn sold to Theodore Lenivitz in 1928. Although the property passed through many hands during the ensuing decades of the twentieth century the mill continued operating. The National Capital Park and Planning Commission acquired the mill for use as a park about 1970.

Story courtesy Michael Dwyer.
Photo Courtesy Montgomery County Historical Society.

Hyattstown Mill
Photo by Brian Lewis
Photo courtesy Michael Dwyer

Laurel Cotton Mill

Maryland

In the early years, mills sprang up along the banks of any stream that promised waterpower. Laurel, no exception, had a gristmill built beside a natural waterfall near what is now the Howard County line, in 1811. Nicholas Snowden leased the mill property to Mr. Johnson in 1824. Johnson converted the gristmill to a cotton mill and set up a factory for spinning cotton yarn. The factory offered employment for approximately one hundred workers. Within ten years, looms were installed and the weaving of cloth began in the mill. History tells us that in 1835, Nicholas Snowden's sons-in-law, Horace Capron and Dr. Theodore Jenkins, took over the property. Some years later, Stephen Heath expanded the activities to include production of spindles, looms, cogwheels, and other related parts. The mill was called the Patuxent Cotton Manufacturing Company and, for obvious reasons, the little mill town called Laurel Factory. Not until June 14, 1870, was the name shortened to Laurel.

To house the families of the mill employees, Horace Capron built a number of two-story brick and fieldstone houses near the mill, each to shelter four families. Rent of one dollar and twenty-five cents to two dollars a month for the dwellings would be deducted from the worker's monthly pay, which started at twenty-five cents a day for a six-day week of eleven hours a day. A few of the original houses are still standing and in good condition.

In 1855, the cotton factory burned. Promptly rebuilt, it stood until the 1930s when it was razed. In the late nineteenth and early twentieth centuries, the mill converted to the manufacture of cotton duck, used for sails and tents. Several decades later it was replaced by the Laurel Swimming Pool. For more information call (301) 725-7975.

Story and photo courtesy of Laurel Museum.

McMahon's Mill

Maryland

The flour milling industry in Western Maryland developed from 1783 to 1820. Milling became one of the earliest industries in Maryland with a good supply of water for power from the many rivers, streams, and creeks. In addition, settlers were moving from the cities to the country. The 1700s gristmills became a necessity to process increasing farm production.

From 1763 to the end of the eighteenth century the most important wheat producing regions of the south centered around Frederick and Hagerstown, in Washington County. (Gray 1933: 608).

McMahon's Mill, built in 1778, had a number of different owners over the years. When the mill acquired a new owner, its name changed to that of the owner. The first documented owner of McMahon's Mill was Benjamin Galloway, a noted local politician. It is said Galloway, a large landowner, had a title, which may have been High Sheriff and that he acted as judge and jury in the area.

Around 1887 McMahon's Mill, one of Shaffer's old Flouring Mills, became known as Shaffer's Mill. Flour, feed, and plaster were ground there. When William Avis became proprietor it became Avis' Mill. Over the past two hundred years McMahon's Mill has also been known as Shanks Mill, Cedar Grove Mill and Charles Mill.

The mill, totally fed by the Downey Branch Spring, is situated on the bank of the Potomac River. In the early 1800s (1828-1850) the C&O (Chesapeake & Ohio) Canal was built along the Potomac River from Washington, D.C. to Cumberland, Maryland. The Canal transported coal, timber, produce and finished products. Because of the location of McMahon's Mill, the owner became involved in canal commerce. In addition the access road to the mill provided the only access to the canal in that area. There is no canal at McMahon's Mill. The eighty-five foot cliffs make it impossible. Boats were taken from the canal to the river and, after a distance of three and one half miles, put back into the canal. The area is known as the "Big Slackwater." A towpath along the river provided a place for the mules to walk and pull the barges.

With development of the railroads and because of storm damage to the canal, the C&O Canal closed to commerce in the 1920s.

McMahon's Mill is a typical four-story mill with millstones located on the second floor. Grain, hoisted to the fourth floor, then gravity fed to the second floor was ground into flour or feed. It would be dropped to the first floor to be bagged or sent back up by belt elevators to the fourth floor to be further ground into different grades. Bags not immediately sold were stored there.

Farmers generally sold the grain to the miller or paid for the milling of the grain with a percentage of the grain processed.

The millstones were driven by power created by the original wooden twenty foot overshot waterwheel. In the 1920s the milling business ended and the owners (Shanks) installed a twenty foot steel waterwheel with a governor and fifteen kilowatt electric generator. Electric power was supplied to the mill and the millhouse.

Around 1976 the Federal Government obtained ownership of McMahon's Mill through the establishment of the C&O National Park. McMahon's Mill was placed on the National Register of Historic Places on August 9, 1979 as part of the park.

Because of their need for waterpower, mills were generally located in flood plains, and once abandoned became victims of floods. Some mills survived floods over hundreds of years. Then new owners removed the heavy milling equipment to use the mill for another purpose. However, because the mill was no longer held on its foundation by heavy milling equipment, it simply floated away.

Story and photo courtesy of William B. McMahon.
Photo by Air Photographics Inc.

Spocott Windmill
Maryland

Cambridge Vicinity, MD
Photo by Ray Tomczak
Courtesy Society for the Preservation of Old Mills

TRENTON MILL

MARYLAND

Trenton Mill in Upperco, Maryland, formerly known as Zouck's Mill, is a gristmill built by George Zouck, circa 1839. According to the local papers, a village called Zoucksville developed around the mill in 1858. Trenton Mill can be found today in a tranquil village that remains much like it was over one hundred years ago. It is believed that Michael Murray Armacost served as the original millwright and in 1861-1862, John C. Zouck built a replacement mill with large stones in each of the four corners.

In 1864, Zouck sold the mill to Mordecai Parrish. Parrish died a year later and was succeeded by tenant operators until John M. Wheeler purchased the mill. At that time, Claudis G. Brown, the miller and considered one of the most valuable citizens of the community felt the dam would not hold. During Wheeler's tenure in June 1866, the twenty-five foot high and 300-foot wide dam gave way .

William Heaffer rented the mill when Wheeler lost the mill by default. In 1868, the town's name changed from Zoucksville to Trenton.

George M. Jordan purchased Trenton Mill in 1920, installed a steel waterwheel and for a while shipped forty tons per month. He sold to McCormack, who sold to Harry S. Morefoot, the last of the line. The dam gave way one final time on July 15, 1951, not to be repaired again.

In 1955, Miss Bea purchased Trenton Mill but only wanted the seventy foot wide stone millkeeper's house which was built in 1836.

She later sold the mill to Sandy Fisher, Jr., a kinsman of the original millwright. He repaired the timber infill at the rear of the building, jacked up the floors, and acquired a Fitz waterwheel. Though much of the equipment is still there, it lacks an abundant water supply.

It is said that one of the last millwrights in the nation visited the mill in 1975 and stated that the framing of the building was typical of mills from Maine to Georgia. Another of his observations after looking at the three forty-eight inch millstones was that the mill had been driven hard. In 1993, John Culver purchased the millkeeper's house and, in 1997, the mill to recombine the properties. Mr. Culver is in the process of renovating and restoring both buildings. For more information call (410)239-9667.

Story and photos courtesy of John Culver.

TROVINGER MILL
MARYLAND

Trovinger Mill, originally known as Jacob Rohrer's Mill, has been situated on the edge of Antietam Creek since the eighteenth century. A date stone on the east gable is dated 1771 with the initials J.B.A. A 1794 map indicates that there were at least ten mills along the creek.

Trovinger Mill, at one time a bustling center of commerce and a local meeting place, is now a quiet reminder of an era past. It is believed that Jacob Rohrer's mill was one of the first mills in operation along the Antietam. Physical evidence suggests mill operation before the date on the stone and that an earlier one-story edifice existed before 1771. This is based on the seams in the brickwork of the building. Patent records from the 1700s substantiate this theory. Jacob Rohrer owned property along the Antietam as early as 1739, and in 1761 Rohrer was granted the patent for a tract of land known as the "Addition to the Jacob Rohrer Mill".

The mill belonged to the Rohrer family until 1817, but changed hands several times over the next five decades before purchased by Joseph Trovinger in 1875. The Trovinger heirs sold in 1910 to Thomas Pitcock.

Mrs. Nellie Lytton, a Hagerstown resident, said she remembers the mill and surrounding areas as "the old homeplace". Her parents moved into the west end of the mill when first married and later to the millhouse. As a young girl she would hang over the half door above the grain chute. When the farmers pulled their wagons under the chute, the children would watch the grain as it dropped into the wagons or visit with neighboring farmers while they waited to get their grain loaded. She said Pitcock owned the mill from 1910-1922.

According to Mrs. Lytton, her funniest memory took place when, as a young girl, she began to climb the staircase and was greeted with a startled shout. About halfway up the steps one of the young men popped his head out the door and said, "Nellie, don't you come any further because we're playing strip poker in here!" *The Maryland Cracker Barrel,* August/September 1994

The mill was a place to visit with friends and catch up on the local news and take care of business at the same time.

Today the mill is owned by Richard and Cheryl (McMahon) Anderson, who are in the process of restoring the historic landmark.

Story courtesy of Richard and Cheryl Anderson. Photo courtesy W.B. McMahon.

Wye Gristmill
Maryland

Wye Gristmill is one of the earliest industrial sites in Maryland. In 1664 Lord Baltimore granted a patent for the land to James Scott, and the first known mill was built on the site in 1682. A gristmill has been grinding here since, over 300 years!

The story of the mill in the seventeenth and eighteenth centuries is a twisted tale of contested wills and inaccurate surveys. It is known that Richard Sweatman enlarged the mill complex from 1682 on, operating two gristmills and a sawmill. Sweatman's milling operations formed a landmark and meeting place so significant that when the Maryland General Assembly met in 1706 to create Queen Anne's County the mill was used as a reference point. Poorly described land patents for the mill led to a dispute over ownership between Edward Lloyd and Richard Sweatman, but in 1722 Lloyd eventually won out. Edward Lloyd (1744-1796), colonial governor, soldier and planter, held title to more than thirty farms, 305 slaves and 11,884 acres of land patented by his great-grandfathers. William Heinsley, of Cloverfield, married Maria Lloyd and acquired the mill property from Edward Lloyd IV in 1778. It was William Hemsley who would have received the purchase order from Robert Morris (financier of the American Revolution and son of Robert Morris of Oxford) for 10,000 pounds sterling worth of supplies for the Continental Army then headquartered at Valley Forge.

The present Wye Mill looks much as it did at the death of William Hemsley in 1825 except for the disappearance of the upright water power sawmill and the busy blacksmith shop. In 1821, Alexander Hemsley sold the mills, blacksmith shops, the pond of thirty acres and tools to Samuel Hopkins for $4,000.00. The Hopkins family operated the mill until 1877 when it was sold to John F.T. Brown, a miller who ground local grains and continued operation of the water powered sawmill until 1899. John S. Sewell bought the mill from Brown and in 1918 sold it to Winthrop W. Blakeslee who operated it until 1953. Blakeslee's descendants are still involved with the mill today. The state of Maryland acquired the mill in 1953 to develop the millpond into a community fishery and flood control project. In 1956 the mill and one acre was deeded by the state to the Society for the Preservation of Maryland Antiquities, also known as Preservation Maryland.

Preservation of the mill has occurred through the help of Preservation Maryland and many local residents and friends. In 1984 Barton McGuire, a noted miller and secretary of the International Society of Millers, began a serious preservation effort. The mill began once again grinding local grains. In 1994 Friends of Wye Mill incorporated and in 1996 Preservation Maryland transferred ownership and operation of the Mill to this local volunteer group.

Today Wye Gristmill grinds Eastern Shore corn and wheat as well as Pennsylvania buckwheat in the traditional stone method using only waterpower. It also houses a museum and gift shop. For more information call (410)827-6909.

Story and photo courtesy Howard McGrane.

Judah Baker Windmill
Massachusetts

In the heyday of the quietly efficient and inexpensive wind power, Yarmouth was once home to five windmills built for the purpose of grinding grain. Over the years, some mills were dismantled for their lumber, some allowed to fall into ruin and others moved to new locations. Since people of that era were quite familiar with overcoming difficult or impossible conditions by using Yankee ingenuity (and with an eye for eliminating undue expenditure), they thought nothing of moving buildings from town to town, often across water. Undaunted, they moved this twenty-one foot by thirty-five foot Judah Baker Windmill several times and Yarmouth's last windmill has the distinction of being one of Cape Cod's "wandering windmills."

Judah Baker's Windmill built in 1791 overlooked Grand Cove in what is now known as the neighboring town of Dennis. Retired sea captains, who were weather experts and kept a watchful eye for sudden changes in the wind, often owned and operated the mill.

It is a smock mill with an unusual conical cap and one dormer. It ground several kinds of grain and, with favorable winds, could grind three bushels of corn per hour. With the advent of steam power, the mill last ground grist about 1891. The interior and exterior of the structure have undergone historical restoration in 1999 with matching funds from the Massachusetts Preservations Project Fund of the Massachusetts Historical Commission.

The mill, built when George Washington served as President, was given to the Town of Yarmouth in 1953. The Judah Baker Windmill now sits on a small, public beach on Bass River, called Windmill Park. It is a contributing structure in the State and National Register of Historic Places in the Bass River/South Yarmouth Historical District. For more information call (508) 398-2231 x 292.

Story and photo courtesy of Deborah Gray, Yarmouth Historical Commission.

Thomas Dexter Gristmill

Massachusetts

The Thomas Dexter Gristmill is one of the best examples of primitive milling in America, preceding the time of Oliver Evans. It was first built circa 1640 and re-built in 1654.

The story begins with Thomas Dexter, Sr., one of the "ten men from Saugus" authorized by the Massachusetts General Court in 1637 to found the town of Sandwich, the oldest town on Cape Cod. As an incentive Dexter was given a land grant of six acres if he would agree to build a mill. Described as one of the most fascinating characters in early Massachusetts history, Dexter was a talented builder of mills, causeways and buildings. He owned a considerable amount of property and held several political positions during the course of his public life. Massachusetts and Plymouth Records indicate that he was frequently involved in litigation, including assault and battery, drunkenness and sedition. In March 1633, he was placed in bilbows, disinfranchised and fined. In 1648 he was involved in no less than six law suits, and although most Yankees considered him to be "a character", he weathered the storms of controversy. Dexter never worked at the mill. He conveyed his rights to his son Thomas Dexter, Jr. and took up residence in the adjacent town of Barnstable. He relocated to Boston in 1673 where he died in 1677.

Thomas Dexter Gristmilll was in continuous operation from 1654 to 1881. Its closing was brought on by a number of factors, including the demise of the Sandwich glass industry, the declining population and antiquated technology. For eighty years the mill remained idle. In the 1950s the citizens of Sandwich became increasingly aware of the historic and aesthetic importance of this "sleeping beauty". A committee of citizens formed to reconstruct the mill and, with great civic pride, completely restored the old mill from its stone foundation to its wooden shake shingled roof. After completion of the restoration in 1962, the mill began grinding cornmeal once again. It is a unique experience to witness an ancient mill in operation grinding corn as done 350 years ago.

The mill is on Lower Shawme Pond in the center of the Sandwich Historic District. It is an idyllic location surrounded by seventeenth and eighteenth century colonial homes. Across the millpond are many beautiful historical landmarks. Among these are a white spired church, modeled after Sir Christopher Wren's architecture, the Hoxie House, said to be the oldest surviving home left on Cape Cod and an ancient cemetery, whose gravestones date back to the 1600s.

The mill, bordered by a stone wall and garden of flowers, and its neighboring town hall, are listed in the Registry of Historic Places.

Even by early mill standards, the Dexter Gristmill is considered small, measuring twenty feet by twenty feet. It is powered by a ten foot cyprus waterwheel. Primitive in its engineering, it operates with wooden gears and old French buhrstones measuring fifty-two inches across.

For more information call (508) 888-5144.

Story and photo courtesy Leo R. Manning, Town Miller.

The Eastham Windmill

Massachusetts

The Eastham Windmill stands as a symbol of a bygone era on Cape Cod. For over 300 years it provided flour, news and gossip for the residents of Cape Cod. Built by Thomas Paine of Eastham, miller, millwright and town clerk, it was originally located in Plymouth, Massachusetts. Sold, it was moved by barge or raft to the Outer Cape sometime before the American Revolution. In 1793 it again sold and was moved overland to Salt Pond in Eastham. About 1808 Seth Knowles, the owner, sold the land it stood on and moved the mill once more to its present site where it operated as a gristmill until 1897. The Village Improvement Society bought the structure for conversion to a library but changed their minds and it remained a mill. In 1928 it sold to the town for preservation. New sails, shingles, arms and wind shaft were installed at that time. The committee who did the work were Harvey Moore, Charles Rogers and William Higgins. Higgins made a temporary lathe in his barn to turn the wind shaft using a "model T" for power.

John Fulcher, who worked in the mill as a boy in the late nineteenth century, was the first modern "miller". Several others have been a part of the mill history and operations since then. Among them are John Higgins the harmonica player, Harold Cole, Freeman Hatch and Jack Webster, and the current millers, Jim Owens and Clyde Eagles.

The 1930 arms were replaced in 1944-1946, 1957 and again in 1962. A steel wind shaft was installed in 1962 and new arms, once more, in 1997-1998.

In 1980 the mill ground corn for its 300th birthday and had a U.S. sub-post office for the summer canceling windmill stamps issued by the U.S. Postal service, which included the Eastham Mill.

Today the mill is a centerpiece for the town and at Christmas is decorated with lights on its arms.

Story and photo courtesy James B. Owens, miller.

MONTAGUE MILL
MASSACHUSETTS

During much of the nineteenth century, twenty-two mills operated on the seven mile stretch of the Sawmill River between Lake Wyola and the Connecticut River in Western Massachusetts. The Montague Mill site consists of a complex of buildings dating from the 1830s, with the main building resting upon the foundation of the original mill built on this site in the mid seventeenth century.

The Montague Mill reflects the changing demographics of the local community and its environs from agrarian to industrial and then to academic/artistic. It first functioned as an all-purpose mill for grinding corn, sawing lumber, and fulling wool for local farmers. In 1834, Alvah Stone purchased the structure. For the next one hundred years, under a succession of owners, operations continued as a gristmill. Grain would be brought to the mill by local farmers and shipped in by railroad. Grinding stones can still be seen along the foundations of the mill at the river edge. The waterwheel pit can be dated from the 1830s, if not earlier.

In 1933 Martin Machine Company purchased the mill and converted it using waterpower to operate a series of metalworking machines in the manufacturing of steel-stamping and cutting machinery for labeling tools, utensils, and baseball bats. A unique stamping machine used to emboss the Louisville Slugger baseball bat has been displayed in the American Baseball Hall of Fame in Cooperstown, New York. The power generating capability of the mill was increased by removing the wooden waterwheel and installing a sixty-three kilowatt electrical generator with two turbines and a penstock, originally used for a street railroad. This equipment and a third turbine is present in the powerhouse today. In the late 1980s, the company moved to a larger facility in Turners Falls and power generation at Montague Mill ceased. The dam has deteriorated and is now breached. A recent owner, Dr. Allen Ross, began an effort to restore the original wooden crib dam and 250-year history of generating power.

During the 1990s, Montague Mill once again changed It now serves the nearby academic communities as a café, bookstore, gift shop, artist studio, and restaurant. The mill complex is on the National Historic Register and has earned a blue ribbon for accessibility from the Boston Society of Architects.

Story courtesy of Mark and Wendy Beaubien.
Photo by Barbara Brady Conn.

Mordecai Lincoln Mill

Massachusetts

In 1691, Mordecai Lincoln, a blacksmith by trade, ended his search for a location that would provide him with accessibility to bog iron ore and a steady flow of water to power a forge. He purchased, for seven pounds, a half interest in a waterfall site on Bound Brook in Scituate, Massachusetts, the boundary line between the Massachusetts Bay and Plymouth Colonies. Although he realized the brook dwindled to a mere trickle in the summer, Lincoln knew that with a touch of ingenuity he could make the site work.

He began by damming the mouth of Bound Brook, holding back enough water each week to power his first mill for two days at a time, a sawmill, leaving the machinery idle for the other five. Hoping to achieve a greater degree of efficiency in his utilization of the brook as a power source, he dammed it a second time upstream and then a third time even farther up. He released the pool on Mondays and Tuesdays to power his iron forge, allowing the water to collect again below. He then moved his operations to his gristmill on Wednesdays and Thursdays, again letting the next dam catch the water. On Fridays and Saturdays he unleashed the flow on his sawmill, from there letting the water flow into the Atlantic Ocean.

Lincoln's inventiveness contributed greatly to the growth of the settlements at North Scituate and Cohasset, providing an invaluable trio of services to the local citizenry. His efforts also paid off financially. The son of an indentured servant, Lincoln died on November 8, 1727 in possession of an estate valued at 3,099 pounds, all from his initial investment of seven pounds. Yet his greatest contribution to American history would come long after his passing.

Lincoln and his wife, Sarah Jones of Hull, Massachusetts, had two sons, Mordecai and Abraham, the latter named for Sarah's father. When old enough, the boys moved to Monmouth County, New Jersey and from there to the Province of Pennsylvania. Mordecai and his wife Hannah had at least five children, one named John, who moved to the Shenandoah Valley in Virginia. One of John's five sons, Abraham, moved to North Carolina, and finally to Beargrass Fort, Kentucky. Abraham had three sons, Mordecai, Josiah, and Thomas. On February 12, 1809, Nancy, the wife of Thomas Lincoln, gave birth to a baby boy in a one-room log cabin on the south fork of Nolin Creek, not too far from Hogdenville, Kentucky. That child, named Abraham, grew up to be the sixteenth President of the United States.

Although currently closed, the original 1691-1692 Mordecai Lincoln sawmill still stands on the roadway named for its builder. Silently nestled in the woods beside Bound Brook, it is a reminder of the intrepid and entrepreneurial men and women who came to this land and with just the simplest of tools built the foundation to the country we call home. For more information, call the Scituate Historical Society at (781) 545-1083.

Story courtesy of John J. Galluzzo.
Photo courtesy Phillips N. Weeks.

Old Mill
Massachusetts

Nantucket Island
Cape Cod , MA
Courtesy Society for the Preservation of Old Mills

Saugus Iron Works

Massachusetts

The Great Migration of the 1630s settled thousands of European fishermen, farmers, and carpenters in the New England wilderness. With these people came the need for traditional iron utensils, tools, and hardware. At the same time, England experienced a shortage of valuable timber used in the form of charcoal to make iron. American iron was needed to help satisfy the demand both in New England and England.

In 1641 "America's first scientist" John Winthrop, Junior sailed back to England from America and assembled the Company of Undertakers for the Iron Works in New England. With investments put forth by the company, Winthrop recruited and shipped skilled ironworkers to Massachusetts. The company established their iron works in Braintree (now Quincy) and in Lynn (now Saugus).

The Saugus Works (1646-1668) included a blast furnace, forge, and rolling mill. The blast furnace separated the iron from locally mined ore. This iron was cast into pots, kettles, skillets, firebacks and salt pans. Iron was also cast into "pig" and "sow,' bars.

Highly skilled ironmakers refined the brittle sow iron in the forge. With a 500-pound waterpowered hammer the iron was worked into wrought iron bars. The rolling mill would further process iron into flat bar and slit rod. The wrought iron was sold by weight to blacksmiths who forged it into a final product. The Saugus workmen passed down special iron making technologies to subsequent generations. They started a manufacturing lineage that continues to the present.

Roland W. Robbins uncovered archeological ruins in explorations from 1948-1953. The museum today houses the remains of seventeenth-century waterwheels and other iron making tools and equipment. Waterpowered machinery from the 1648 saw blade and edge tool shop of blacksmith Joseph Jenks is also on display.

A complete reconstruction of the iron making buildings opened in 1954. During the summer, seven waterwheels, four pairs of bellows, a five hundred pound hammer, and rolling and slitting machinery provide a unique opportunity to experience seventeenth-century iron making technology first hand.

For more information call (781) 941-2372.

Story courtesy of National Park Service
Curtis White, Park Ranger.

Souther Tide Mills

Massachussets

The Souther Tide Mills in Quincy, Massachusetts are among America's last surviving coastal tide mills. Prior to the nineteeth century, wind and waterpowers were used extensively in America. Where the hilly land provided rivers with waterfalls, conventional river mills were erected. In coastal locations, however, in flat land, river mills could not be built. There, tide mills were constructed. The general method used to extract energy from ocean tidal power required the construction of a dam across a tidal inlet or river mouth, and a set of one way gates in the dam. When the tide would rise, all the water in the area would fill. When the tide went out, the gates shut by the water's force, and the miller had an eight or ten foot deep pond of salt water with which he could run a mill. Tide-millers had to work at different hours every day, yet they had the advantage of a never-ending power source located close to major shipping channels. Consequently, many of these mills were built.

In the six New England states and New York, covering the Atlantic coast from Maine to New Jersey, over 150 tide mills operated between 1631 and the 1930s. Over ninety percent have been lost over the years, victims of fires, floods, urban development and other causes. Currently, there are thought to be less than ten tide mills surviving in the entire Northeast United States.

Among the finest are the Souther Tide Mills of Quincy, south of Boston. These mills include a circa 1802 gristmill, a circa 1880 sawmill, and a connecting lumber mill structure adjacent to an 1815-1851 wooden shipyard site. The mills and shipyard site are located at the mouth of the Town River, which also functioned as Quincy Canal and Quincy Granite stone export center between the 1820s and 1920s.

The site is currently privately owned, not yet restored or open. However, the property has been declared eligible for listing on the National Register of Historic Places. The City of Quincy and many non-profit agencies are working to preserve and redevelop the Souther Tide Mills as an historic site and Waterfront Heritage Park. It is anticipated the park will open in or shortly after the year 2000, with the Tide Mills repaired and used as a museum. Plans also call for the site to maintain tall ships and water tours of nearby Town River Bay, and Boston Harbor. The mills are in Quincy, close to the birthplaces of U.S. Presidents John and John Quincy Adams.

Story and photo courtesy John Goff.

Stockbridge Mill

Massachusetts

Isaac Steadman arrived in the New World aboard the Elizabeth in 1635. He came not in search of religious freedom or political sanctuary; instead he left his native Biddenden, Kent, England on an opportunistic journey toward financial independence. Just north of the settlement of Plymouth he located a small but active community of kindred souls, other men of Kent, residing in a village known as Greene Bush.

In 1640, Steadman dammed First Herring Brook at Greene Bush, and built the first waterpowered gristmill in the old colony. On June 7, 1648 he was admitted as a freeman and he became a constable of Scituate, Massachusetts. Two years later he sold the mill to George Russell of Hingham.

Russell operated the site for six years before becoming partners with the notoriously vociferous anti-government John Stockbridge, who operated the mill for less than a year before moving to Boston. Stockbridge left his share of the mill to his son Charles, who bought out Russell in 1665.

On May 20, 1676, the mill nearly fell to the torches of King Philip's raiding Indians. Under orders from their leader to burn all the villages of white settlers, they first set fire to Russell's and Stockbridge's sawmill, just to the west. After setting the surrounding houses ablaze, they advanced on the gristmill. There the valiant men of Scituate made their stand, driving the marauders out of the town by the end of the day.

For the next 154 years members of the Stockbridge family ran the mill and then sold to the Clapp family, who operated it until the turn of the twentieth century. Realizing the economic inefficiency of the mill in an age of rapid industrialization and astounding technological improvement, the Clapps shut down the site in 1900. In 1922, they donated the still operable mill to the Scituate Historical Society.

Over the past seventy-seven years, the Society has occasionally opened the mill and the gate across the roadway at Old Oaken Bucket Pond, setting the machinery in motion to grind corn. Although experimentally powering it by electricity in 1961, the Historical Society called in professional millers in 1970 to restore the mill to run by waterpower, and later to sharpen the stones. It last ground corn in the 1980s.

The Scituate Historical Society periodically opens the Stockbridge Mill and although the mill is not currently in operable condition, that may change in the very near future. For more information call (781) 545-1083).

Story and photo courtesy of John Galluzzo.

Sturbridge Village Gristmill
Massachusetts

(aka Wight Gristmill)
Old Sturbridge Village (Worcester County), MA
Courtesy Society for the Preservation of Old Mills

Sudbury Gristmill
Massachusetts

Longfellow's Wayside Inn
South Sudbury, MA
Courtesy Society for the Preservation of Old Mills

The Atlas Gristmill

Michigan

Built in 1835, the Atlas Gristmill in Flint, Michigan originally sat on the Kearsley Creek Millpond in the Village of Atlas. The mill, operational until 1942, is the oldest known mill in Michigan.

In 1975, the mill gained designation as a state historic site. The Genesee County Parks and Recreation Commission also purchased it the same year. The mill was moved to Crossroads Village and the Huckleberry Railroad, a living history village and steam railroad just north of Flint, Michigan. Although relocated, the mill has not lost the historical designation.

The builder of the mill, Norman Davison became the second settler in Atlas Township. He moved from New York to Michigan in 1831. When Davison and two of his seven children arrived the area was still a wilderness. He built a house and then dam on Kearsley Creek where he first operated a sawmill. After he built the gristmill, the settlement was named Davisonville and the post office operated from the mill. Davison served as the first postmaster and later served as a township judge and a member of the convention that wrote Michigan's first constitution.

The three-story frame building, originally much larger, also housed a woolen mill. A belfry contained a bell used for fire alarms. The grindstones were powered by water flowing through a chute to a boxed wooden turbine. In the 1860s, water driven turbines replaced the inefficient wooden turbines twelve feet below the stream surface. This enabled the mill to operate through the winter season, unlike mills utilizing waterwheels that froze and were not functional for three to four months.

The mill was moved to Crossroads Village in October of 1975. It had to be cut in two horizontally for the twenty mile trip. A millpond constructed on the village site and the mill completely restored, it reopened in 1976. It continues in operation as part of the interpretive program at Crossroads Village and is the most impressive of the collection of over thirty historic structures. For more information call (800) 648-7275.

Story and photo by Janet S. Weaver.

Bellevue Mill

Michigan

The Bellevue Mill is on Battle Creek in the village of Bellevue, Michigan. Isaac Crary and John Pierce constructed the first dam and sawmill on the site in 1834. In 1842 Issac and his brother John erected the first gristmill with two runs of stone. In 1854 Manlius Mann bought the gristmill site and, with Horatio Hall as carpenter, built the present structure. It was equipped with a ten-foot diameter overshot waterwheel which was enclosed in the basement to help prevent icing in winter. Grinding capacity was upgraded with three runs of stone. Mr. Mann married the local Potawatomi Indian chief's daughter. Members of the tribe assembled the upper structure of the five-story mill. It seems the Indians did not mind the height as much as the white men. Another local story tells of some boys wanting to play a prank on Mr. Hall, the carpenter. They cut several inches off the end of his master measuring pole thinking it would cause him to cut the beams too short. Horatio had the last laugh because he used the middle portion of the pole for measurement, keeping a short length at both ends for wear.

The single story on the west end of the building was added about 1871. The mill, converted to a roller process in 1888, increased flour production to seventy-five barrels per day. Later, the substitution of two forty-five horsepower Leffel turbines for the waterwheel increased production to 120 barrels per day.

By 1881 the ownership of the mill had passed to Hiram Ovenshire and remained in his family for the next four generations ending with the Butler brothers, Bob and Chuck. Bob took over mill operations in 1937, at the age of eighteen. In 1958, due to changing economic conditions, the mill closed. The equipment was sold and removed for use in Mexico. The empty building became subject to vandalism and deterioration until 1977 when purchased by William and Carole Jean Stockhausen, the present owners. Just before the purchase, members of the Bellevue Historical Society managed to have the mill placed on the National Register of Historic Places. The Stockhausens decided to restore the building to its original configuration, so the 1871 addition was removed and new doors, windows, siding and roof were installed. By 1983 they had also repaired the civil works and rebuilt and installed salvaged turbines and induction generators. Using the flowage from Battle Creek the mill now generates just over 200 thousand kilowatt hours of electricity per year. The Stockhausen family plans to restore the interior for use as a home.

Story and photo courtesy of William Stockhausen.

Bellevue Mill

W.H. King Flour and Feed Mill

Michigan

Almont (Lapeer County) MI
Photo By Palmer Werner
Courtesy Society for the Preservation of Old Mills

KING MILLING COMPANY

MICHIGAN

In 1890 the Superior Mill of Lowell, Michigan filed for bankruptcy. A local lumber concern of Francis King, his son, Frank T. King, Charles McCarty and Reuben Quick purchased the property for $20,000 at a mortgage sale. The King Milling Company began. The old Superior Mill converted to the new "roller" process of grinding grain with an increased capacity of 500 cwt. per day. In 1896 the company purchased the Forest Mills, also in Lowell, and converted it to corn milling.

In 1900, Thomas F. Doyle, a successful lumberman and farmer, moved to Lowell from the Irish community of Parnell and purchased stock in King Milling Company. In the next few years, his three sons, Charles, Renis and William C. (Bill) went to work there.

In 1927, the company acquired Peckham Furniture Company, at one time the largest Cutter manufacturer in the world. This property had waterpower. William Doyle converted the Peckham building into a poultry and livestock feed manufacturing facility. From that part of King Milling Company, Doyle founded the Michigan Feed Manufacturers Association in 1933 (now the Michigan Agri-Business Association) and became the first president.

Francis King, the founding president of the King Milling Company served in that capacity until his death in 1900. Frank T. King followed his father as president and served until his death in 1934, the last decade as consultant to the three Doyle brothers who actually managed the company.

Charles Doyle followed Frank King as president and remained as such until his death in 1943 although inactive because of a paralytic stroke in 1936. His brother, William, then assumed management of the mill.

When William Doyle died in 1945 his majority interest passed to his invalid wife, Monica, his twenty-three year old son, Naval Ensign King, and his fifteen-year-old son, Roland (Mike). The Navy released King Doyle from active duty because of his father's death and he became president of the company, a title he held for fifty years.

The family of Charles Doyle sold their interest to King Doyle in the 1940s and the Renis Doyle family sold to the company in the 1950s.

Under the direction of King and Mike the company has grown from 600 cwt. of white flour in a twenty-four hour day with storage of 80,000 bushels, to its 1990 capacity of 5,400 cwt. of white flour, 5,000 cwt. of whole wheat flour and storage of 2,800,000 bushels. The production method was changed from bucket elevator conveying to pneumatics, before popular in the milling industry.

The company is now completely automated from its grain storage to the finished product bins, and primarily grinds Michigan soft white wheat developed by Michigan State University.

Currently, King and Mike are less active in the day-to-day operation of the mill and King Milling Company is being managed by the fourth generation of the Doyle family. King's son, Brian, is President. Mike's son, Jim, is Senior Vice-President and Mike's other son, Stephen, is Vice President.

In the year 1890 there were 700 flour mills in the State of Michigan. Today there are only six.

Story and photo courtesy King Milling Company

Millhouse
Michigan

(Wolf Lake State Fish Hatchery)
Mattawan (Van Buren County), MI
Photo by Palmer Werner

Old Cartwright-Van Patton Mill

Michigan

Homer (Calhoun County), MI
Photo by Palmer Werner

The Old Mill

Michigan

The Old Mill, built in the Greek Revival style in the 1850s, sits beside the River Raisin in Dundee and is surrounded by park, picnic, playground and riverwalk areas. In 1911 mill owner, R.B. Davis, sold to the Dundee Hydraulic Power Company. The new owner replaced a broken wood rafter dam with a concrete dam. A lawsuit instituted against the new dam claimed it to be higher than the old one. After losing in both the local court and the state court, the new dam had to be lowered. The lawsuits with their testimony, pictures and diagrams provided historical information that otherwise would have been lost; but they also served to inhibit Henry Ford's plans some twenty years later. For about ten years the two seventy horsepower Trump turbines in the mill powered an electric generator for the village and machinery for feed mill processing. When the electric lines from Detroit Edison reached Dundee, the mill, abandoned, became an eyesore and a fire hazard.

In 1931 the Village Council embarked on an improvement project, with plans to raze the mill. When the state fire marshal contacted the owner of record he learned the new owner was Henry Ford. In the early 1920s several locals had tried to interest Mr. Ford in making the mill part of his Village Industries project. After much speculation on the part of the local citizens, Mr. Ford's work crews began reconstructing The Old Mill in 1935. He restored the three-story mill to its original appearance with extensive interior improvements and placed machine tools on each of the three floors. On the river side of the mill he built a limestone addition, using stone blasted out of the river bed for the new tailrace. The limestone addition housed two GE generators, one steam powered and one waterpowered by a Leffel Turbine with a concrete draft tube to the tailrace, two large boilers and a foundry. The Old Mill remained a Ford Village Industry until 1954. Ford's Mill melted scrap brass from the large plants on the Rouge, cast them into welding points, and did the finishing work. The project provided a major boost to the local economy during the Depression.

In 1954 Ford Motor Company sold to Wolverine Manufacturing Company, who built two additions for a warehouse and offices and converted it to a paper mill for manufacturing gasket material.

In 1970 Wolverine sold the landmark to the Village of Dundee for $1.00. Since then, under the auspices of the Village Council and the Old Mill Restoration Committee consisting of a group of local volunteers, the Old Mill has undergone extensive remodeling to serve as a Community Center (warehouse) with kitchen facilities, council chambers (office) and Museum (Old Mill).

The Old Mill, part of the Ford Heritage Trails, is on the National Register of Historical Places. For more information call (734) 529-3038.

Story and photo courtesy Douglas & Dorothy Heinlen.

Parker Mill

Michigan

Parker Mill County Park encompasses forty-five acres of the original Parker family homestead of 1863. The county-owned portion is a twenty-seven acre tract that includes the historic Parker gristmill complex, rich Fleming Creek forest lands, the old millpond site, and a small remnant of prairie-fen habitat. Situated between Ann Arbor and Ypsilanti, Parker Mill is the only operational pioneer gristmill remaining in Washtenaw County.

William Quainton Parker, a millwright, and his new bride, Mary, came to Washtenaw County in 1861 from Quainton, England. In 1863, they established a farm at this site on sixty-one acres purchased by Mary using money saved from an ale allowance she received as a lady's maid in England. On the ruins of the old Fleming sawmill of 1824, William constructed a small gristmill in 1873 to produce corn feed for the family farm. In 1910, after William's death, his son George took over the operation of the mill, expanding it to commercial flour production. George's older brother, Fred, was a carpenter and laborer instrumental in the construction of the adjacent cider building. Two milling stones, made of French quartzite brought over as ship's ballast, allowed for the simultaneous production of corn and wheat products. George Parker produced wheat flour, pancake mix, graham flour, buckwheat flour, corn meal, and cracked wheat breakfast cereal. Milling ceased in 1959, three years after George Parker's death. The Parker's descendants remain in the region and continue to provide historical insights.

Like many pioneer structures, the Parker Mill buildings were built mostly of materials gathered on the farm, glacial field stones, riverbed sands and gravel, and bottomland hardwoods. Rather than the familiar waterwheel of other mills, an indoor turbine in the lowest level of the gristmill harnessed the power of Fleming Creek. Water from the former millpond north of Geddes Road entered the mill through an underground tunnel. The main driveshaft transferred the power from the turbine to a series of gears, shafts, pulleys, and belts where it could be used to turn the milling stones, elevators, cleaners, and sifters. Additional drive belts were added to run the equipment in the cider mill, as well as a small lathe in the gristmill. A 1978 flood destroyed the millpond and dam. An electric motor now powers the milling equipment.

Washtenaw County Parks and Recreation Commission purchased the mill in 1983 and restored it to the original working condition. Parker Mill is a natural area reserve and historical education facility. For more information call Washtenaw County Parks and Recreation Commission at (734) 971-6337.

Story and photos furnished by Matthew C. Heumann.

Peterson Mill
Michigan

Clifton Wade built the first mill on this site in 1814 and used it to power a tanning process. In 1868, a spring flood destroyed this mill.

Rolland E. Peterson constructed the present mill (Peterson Mill) in 1959. The undershot wheel is twelve feet in diameter with wider than usual blade spacing. Moores Creek provides the stream that powers the mill. The stream, about five miles in length starts from the overflow of nearby Lake Goshorn.

Peterson Mill powers a small set of stones, primarily used for corn and buckwheat. The fresh cornmeal has been used for years to make very popular corncakes.

The mill is operational in the summer but closes for the winter so the millpond can be drained. For more information call (616) 857-2151.

Story courtesy R.J. Peterson.
Photo courtesy Society for the Preservation of Old Mills.
Photo by Palmer Werner.

Peterson Mill
Conyers (Allegan County), MI
Photo by Palmer Werner

SCOTT'S MILL
MICHIGAN

Scott's Mill is located in Kalamazoo County, Michigan in Kalamazoo County Park. The oak waterwheel is eleven feet in diameter. This is the source of power for the machinery required to grind grain into flour and corn into meal.

Records indicate that the gristmill, valued at approximately $5,000, was built in 1867 by Servitus Bathrick, a Baptist minister and practicing physician. The waterwheel produced forty-seven horsepower.

Miller E. V. Ford ran the operation alone and earned $250 for five months work. Bathrick reported inventory of 7,800 bushels of wheat, corn, oats and feed grain. The inventory also reflected $6,400 worth of flour and feed for the year.

The mill sold at public auction in 1876 to Nelson Eldred. Eldred then sold the mill to William White in 1879. In 1880 the sawmill and gristmill were still operating. White ran the gristmill year round with one helper. He had three runs of stone and a three-foot wide "Lassell" wheel turning twice per second to produce about twenty-five horsepower from a fourteen foot head of water.

In 1884, according to the census taker, White's sons William and George (age fourteen and twelve) were the millers. Charles (another son) joined four years later.

The mill was overhauled in 1893 and the latest roller mill equipment installed.

In 1974 an anonymous owner donated the funds for a report on the feasibility of restoring and developing the mill and agreed to back the restoration of the mill and miller's home.

From late fall 1974 until September 1976, members of the Kalamazoo County Parks and Recreation Department spent part of their time working on restoration of the mill and constructing other facilities. Repairs were made to the dam and on Oct 8, 1976 Scott's Mill was dedicated as a public park. The mill is also a State Historical Site.

The restored mill is used to illustrate how facilities of this type operated in the past. Old Mill News Spring 1991.

Courtesy Society for Preservation of Old Mills.

Scotts Milling
Kalamazoo County, MI
Photo by Palmer Werner

Pears Flour Mill

Michigan

Nineteenth century Buchanan, Michigan was a town of mills. At least thirteen water and steam powered mills were counted along the banks of McCoy Creek, the stream that flows through the center of town. The Pears Flour Mill, built in 1857, is the only one of those mills still standing.

Eastern investors financed the Pears Mill, originally known simply as The Buchanan Mills, as a state-of-the-art replacement for a smaller mill already occupying the site. By 1860, the three-story Mill with a sixteen-foot head of water and four runs of stones, each five feet in diameter, was the largest and most profitable industry in the county and a local player in international trade. A downturn in the flour trade, the introduction of new milling technology and concentration of the wheat and flour industry farther west led to a decline of fortune. By 1886 the Mill closed. During the next twenty-five years the top story was removed and the mill refurbished, reopened, and operated as a feed mill and custom gristmill for local farmers. Operations continued under the Buchanan Co-ops until 1983, when the mill again closed and sold to the city's Downtown Development Authority.

Initial plans called for the demolition of the Mill, but through the concerted efforts of a local historical group, the Buchanan Preservation Society, the community decided to save the building. All modern additions were to be removed and the mill restored as an operating nineteenth century mill. In honor of William and Charles Pears, two longtime owner/operators of the Mill, the newly saved structure was renamed Pears Mill. Restoration work began in 1984 and the walls and first floor completed in 1985. Antique machinery salvaged from an abandoned mill in 1988 was installed and in 1989 the doors of the mill opened.

Currently, the ten foot wooden waterwheel powers one run of forty-two inch stones. The water that powers the wheel flows through the original half-mile long millrace built in 1848. Future plans call for installation of line shafts, a variety of antique belt driven machinery and further restoration. For more information call (616) 695-5525 or contact the Buchanan Area Chamber of Commerce.

Story and photo courtesy of Peter J. Lysy.

Photo taken from border of large wall map published in 1860.

Water Powered Electric Millhouse
Michigan

Dearborn (Greenfield Village County), MI
Photo by Palmer Werner

YARMOUTH WINDMILL
MICHIGAN

Cape Cod Windmill
Dearborn (Greenfield Village), MI
Courtesy Society for the Preservation of Old Mills

Yarmouth Windmill
1633 Cape Cod Smock Windmill
Dearborn (Greenfield Village), MI
Photo by Palmer Werner

Wolcott Mill

Michigan

Along the North Branch of the Clinton River in rural Ray, Michigan, the Wolcott Mill stands tall and proud in an idyllic setting. In the final curve of the quiet road the view of the mill is breathtaking during any season.

Wolcott Mill is a three story gristmill and monument to the agricultural heritage. The austere Greek Revival architecture of the mill symbolizes the values of the past as an icon to the disappearing farming community.

Built in 1847 by Arad Freeman, and originally named Monterey Mills, the mill was once the hub of the community. Farmers needed a gristmill to help process their grain crops, and the mill had to be relatively close because the heavy loads could only be transported short distances with horse drawn carts. The mill served a social function as a place to hear news of the community and purchase items needed for the homestead. Serving local farmers for 120 years, the mill at one time produced an average of 40,000 pounds of buckwheat flour and 30,000 pounds of meal annually. The Wolcott Mill became well known for silk pastry flour.

The mill changed owners and names many times, until 1878. At that time Frederick Beech Wolcott purchased it. Leaving their homestead in Nebraska, Fred and his wife, Anna, moved to Michigan and began milling.

Finding the mill in a state of disrepair, Fred immediately started updating the necessary equipment. In 1881, he replaced the original fourteen foot wide, sixteen foot diameter waterwheel with a thirty and one half inch Leffel turbine. In 1890, three sets of forty-eight inch grinding stones were replaced with a longer system of roller mills. In 1917, an even more efficient system of Midget Marvel short roller mills was installed, and the single turbine was replaced with a seventeen and a half inch and a twenty-three and a half inch Leffel turbine. Fred and Anna had twelve children, and the Wolcott family owned and operated the mill until it closed in 1967. For more information call (810) 749-5997.

Story by Rachel Godley, History Interpreter.
Photo by Nan Monroe

Phelps Mill

Minnesota

The history of milling in rural Minnesota is embodied in the story of Phelps Mill. By the late 1800s nearly 1,000 mills were operating throughout the state.

Otter Tail County was considered a prime location for the construction of mills. An abundance of waterpower lured entrepreneurs with dreams of turning the county into the largest flour producing area west of Minneapolis.

One such man was William Thomas who owned and operated a flour and feed business in Fergus Fails. In January 1887, Thomas purchased thirty seven acres in Section 33 of Maine Township. Running through this land was the Red River. Thomas considered this a perfect location for a flourmill. He sold his business and with his wife Nonie, moved to a log cabin above the spot on which he planned to build his mill.

In the spring of 1888 he began work on the dam. The first dam constructed of wood, had an irritating tendency to leak. Workers often had to use sandbags and loads of gravel, dirt, manure, hay and straw to plug the steady stream of water gushing through to the other side.

The mill was constructed of wood of which Thomas paid two cents a foot for pilings and three cents a foot for sawn logs and square timber. According to an April 1889 Fergus Falls Weekly Journal article, the thiry-six foot by thirty foot mill was framed by Royal Powers without the aid of a blue print or sketch. Though he had no construction diagram to work from, Powers put the building together without marking a stick of lumber - keeping the whole plan in his head.

The finest machinery, costing close to $5,000, furnished the mill. The turbines weighed over 7,000 pounds and were transported from Underwood by horse and wagon. It took a full day to make the nine-mile journey.

Construction was completed in October 1889, in December the mill began operating. Known as the Maine Roller Mills, it would produce sixty to seventy-five barrels of flour per day. It made patent, straight, bakers and low-grade flours.

The mill met with considerable success, for after a few months of opening the Maine Township Board had its hands full laying out new roads to the mill. At the height of the wheat grinding season, twenty to thirty-five wagons loaded with sacks of wheat would line up outside the mill. Farmers from a distance stayed overnight in the "Farmers Roost," a bunkhouse Thomas constructed north of the mill. He also provided a barn where horses could be stabled free of charge.

In 1895, after grinding more than 44,000 bushels of wheat and 25,000 bushels of feed the previous year, Thomas decided he needed more machinery and space. As a result, he constructed a twenty-five by thirty-six foot addition on the north side of the mill. Covered with sheet iron, the addition was used for the grinding of buckwheat and rye.

After the turn of the century, business at the mill gradually declined. Steam, gasoline and electricity proved to power mills more efficiently than water. In addition, railroads made it cheaper to ship wheat than to mill the wheat in the county. Soon rural mills became obsolete.

Thomas sold the mill in 1919 to the Farmers Mercantile Company of Underwood for $35,000. They could never make a go of it as many customers complained their flour was not up to the mills old standards. They sold out in 1928 to H.G. Evenson of Wall Lake who offered $10,500 for the purchase of the property. Evenson ground grain for stock feeding on a limited basis. The mill finally closed its doors in 1939. For more information call (218) 736-6038.

Story and photo courtesy Chris Schuelke.

Pickwick Mill
Minnesota

In the 1850s George W.T. Grant first visited the Big Trout Creek area. In 1854, Grant recognized that the falls would be an excellent waterpower source and decided to construct a gristmill.

In 1856, Timothy and Wilson Davis, became associated with Grant and built Pickwick Mill. The current structure, completed in 1858, is six stories and sixty by forty-five feet, constructed of limestone (quarried locally) and built without nails or pegs. The limestone walls are over three feet thick at the base and taper to twelve inches at the peak.

Pickwick Mill ground flour until 1919, as much as one hundred barrels daily. Local farmers hauled the flour two miles to the Mississippi River to what was known as McGilluery's Landing and Ferry. There it was shipped to the eastern United States, Europe and South America. After 1919, the mill continued grinding feed for the local farmers.

Pickwick Mill still contains nineteenth century flour milling machinery, representing the technological changes that took place during the 1900s.

On September 20, 1980, a flash flood turned Trout Creek into a raging monster that tore through the village of Pickwick. During the flood, the dam and spillway were destroyed and water rose halfway up the walls of the third floor of the mill. Only the thick stone walls and strong construction saved it. The elderly owners then decided to close the business.

When talk of razing the building began in 1982, concerned citizens formed Pickwick Mill, Inc. and purchased the mill. Thus, began the tremendous task of raising money to restore the dam, pond gates, and building.

Since the purchase in 1982, renovations include replacement of the roof, beams, flooring, and most of the windows. Original gates were restored and a new safety dam built. The millpond, "Lake LaBelle", also had to be rebuilt. A new twenty foot by four foot waterwheel replaced the turbine installed in the 1880s and generated the power to run the new millstones.

Pickwick Mill, Inc. is striving to become a "milling interpretive center."

Story by Opal Fitch.
Photo by Kay Shaw Photography.

SCHECH'S WATERPOWERED MILL

MINNESOTA

Schech's Waterpowered Mill was built of limestone rock in 1876 by John Blinn. Few mills existed in Houston County in the 1800s, most were made of wood and are now gone. Blinn called his mill the Caledonia Mill. After John Blinn's death in 1885 M.J. Schech bought it.

Schech, head miller of the St. Paul Mill, brought steel rollers and purifiers from the cities to upgrade the process of milling flour. He then changed the name to The Caledonia Roller Mill. The mill produced any type of flour known and cereal. One of his two sons, Edward, took over the Mill about 1912 when Schech retired. The power used to run the mill consisted of three water turbines, new in 1924.

Ivan Krugmire and his wife, Eleanor, the daughter of Ed Schech, have been maintaining and using the mill since 1946. Waterpowered mills require significant maintenance for the Dam of Turbines and for that reason few still use waterpower. The first wooden dam was replaced replaced with a cement dam in 1922. Flash boards on top held in place with wire keys, prevent damage or destruction from water pressure of floods. Each fall during flooding, mud settles in the wheel pit and the Krugmires must climb into the pit to remove the deposits. Schech's mill has been in the National Register of Historic Places since 1978.

Story and photo courtesy of Ivan Krugmire.

TERRACE MILL

MINNESOTA

The Terrace Mill Historic District portrays the era when Minnesota was known around the globe for vast production of fine flour. Located in the picturesque hamlet of Terrace, the existing mill, constructed in 1903, is the third structure on this site since 1870. The structure remains essentially as it was when in operation. The Terrace Mill is the only remaining of nine in Pope County, and one of only a few of nearly 1000 in Minnesota that operated around the turn of the century.

The Leffel turbine that powered the mill is still in place. A headrace opening is under a dock in the mill pond, although sandbagged and awaiting the relining of the flume leading to the turbine. When the mill was converted to a furniture manufacturing factory much of the equipment was destroyed. Replacement pulleys, gears, shafts and belts are being sought with the goal of making the milling equipment operational.

The Terrace Mill Foundation purchased the mill in 1979 when it was placed on the National Register of Historic Places. In addition to restoration of the milling mechanism, the Foundation has developed the Terrace Mill as a center for various cultural, ethnic and artistic activities. The third floor houses collections of original art, photography and works of literature. The second floor is used as a theater for various productions and music performances. All floors also contain a vast collection of artifacts including trunks, tools, cabinets, looms, dolls, etc.

In addition to the Terrace Mill, the Historic District includes a fully restored and equipped log cabin and Norwegian Cottage, a Keystone Arch Bridge reconstructed in 1998, and the beautiful Chippewa Falls, all located within a few yards of the Mill. For addition information call 320-278-3728.

Story courtesy of James D. Torbert, Vice President, Terrace Mill Foundation.

Cave Spring Mill

Mississippi

Enterprise (Clarke County), MS
Photo by J.C. Barger

Sciple's Water Mill

Mississippi

Dr. Hunnelly built the original water mill on this site in 1790. He and many others migrated to Kemper County from Watree River Country of Kirshow County, South Carolina. This information comes from Col. Jim Walton who was born around 1750 and lived in Kemper County for many years. He was a recognized authority on Indian lore and early oral history of Kemper County.

During the next forty to fifty years, four families owned and operated the mill. The Sciple Family purchased the mill from Mr. Footes and Rose McManis between 1830 to 1850. They have owned it since that time and seven generations have worked there.

Sciple's Water Mill has been in operation continuously since built with the exception of three months, when a tornado destroyed the mill, November 28, 1973. Many people provided generous help to rebuild on the original foundation. The floor and foundation were undisturbed by the storm. On March 4, 1974, operations began once more.

Until the mid 1950s the mill ginned twenty-five to thirty bales of cotton and sawed 10,000 board feet of lumber per day. Originally a two story structure, the cotton gins were on the second floor and the sawmill under the shed over the creek.

The mill is powered by a Leffel water turbine wheel. A wood turbine, used until 1880 provided less horsepower.

Water is supplied by springs running from the hills upstream. The millpond covers twenty-five to thirty acres.

Story by Ed Sciple.
Photo by Land Photography.

Dawt Mill

Missouri

In 1892 John C. Caldwell purchased the property and mill. L.A. Rogers designed the dam and millrace and constructed a 425-foot long angling dam. For construction of the dam, sycamore timbers and rocks were used. Tragically, Rogers drowned in the millpond in 1893. A short time later, the log structured mill burned.

Between 1897 and 1900 Alva Hodgson, builder of Hodgson Mill at Sycamore, purchased the land. The site was an excellent location and had a dam. On January 12, 1901, Alva registered the purchase at the Ozark County Courthouse. Records show Alva, his brother George, and his mother, Mary E. as the owners.

Alva rebuilt the dam, improving it by using stone anchored to submerged oak trees and capped with heavy boards. This acted as a spillway by directing the current up and over the dam to channel it into the millrace. A three-story roller mill replaced the previous one. Although farmers changed from raising grain to cattle and most mills were no longer needed, Dawt Mill remained operational.

Charles Thompson sold the mill in 1945 to Ray Hunt. Hunt and his wife ground feed for cattle, added gas pumps and installed new French stone buhrs, still used today. One of the old buhrs is now in the foundation wall of the porch of the mill.

In the 1950s Karl Schmidt built the overshot waterwheel located on the north end of the mill for aesthetic purposes only.

Ray Hunt died in 1964 and his wife continued to operate the mill until purchased by a group from Mountain Home, Arkansas.

Many floods have affected the mill, two in 1905 and 1918. In 1968 large sections of the dam washed out. In the summer of 1969 the Dinnell's, then owners, and several other property owners in the area financed the rebuilding of the dam. Waters from a flood in 1982 completely covered the millrace and the bridge by more than eight feet. In the spring of 1990 the bridge and dam were under water. Houseboats moored by the bridge and boats went upriver over the dam. In April 1992 a flood took out a section of dam. Repairs were made in March 1993. Floods inundated the Midwest in 1993 but the area around Dawt remained dry until September when heavy rains in a forty-eight hour period caused major flooding and damaged the support beams. In October a brick and concrete block wall was built under the mill. In November the floods hit with even higher waters. The mill and dam remained intact but suffered from water damage.

The mill has stood for nearly one hundred years in spite of natures severe blows, and serves as a testimony to Hodgson's ability as a builder.

Dawt Mill is the only one still operational in the area. The grinding of cornmeal and whole wheat flour is done by waterpower and stone buhrs. There is a gift shop on the site.

Story and photo courtesy of Cindy Roberts.

At the turn of the century Dawt had a grocery store, post office, blacksmith shop, saw mill, cotton gin, roller mill, a few houses and a church. The mill became the place to meet with neighbors while waiting to get corn and wheat ground, or cotton ginned. As early as 1866 a mill had been established on the property.

In 1874 President Grant conveyed a patent for the land to Ruhama J. Isom, who homesteaded the property prior to that date. The land sold the same year to Samuel S. Price. The following year Price sold to James Colson due to an order to pay alimony of $500 and court costs for the divorce of his wife, Mary A. H. Price. Shortly thereafter Colson sold to Ben B. Price and John H. King. At the time, the deed had been lost or destroyed and not recorded. In February 1878 the property sold at a sheriff's sale for the $500 debt plus $17.35 for damages.

Bollinger Mill

Missouri

A rare combination of a nineteenth century waterpowered gristmill and a 140-foot wooden covered bridge make Bollinger Mill State Historic Site a unique opportunity to envision the past. The mill offers the sights and sounds of grinding corn into meal with waterpowered millstones and demonstrates the many types of machines and processes used to make flour from wheat. These two National Register structures provide a wonderful backdrop to history.

George Frederick Bollinger left his home in North Carolina in 1797 and met Louis Lorimier, the Spanish commandant for the region, in Cape Girardeau. Lorimier promised a land grant to Bollinger if he would help settle the area. Bollinger returned to North Carolina and on New Years Day, 1800, he crossed the Mississippi River with twenty families. They formed the Whitewater settlement and Bollinger built his first mill. Bollinger replaced the first mill in 1825 with a more substantial structure. His new mill had a limestone foundation with a wooden upper story.

When Bollinger died in 1842, his only child, Sarah Daugherty, inherited all of his property. At the beginning of the Civil War one of Sarah's sons was accused of firing at a Union wagon train coming through the area. In September of 1861 Union troops burned the mill in retaliation.

Solomon Richard Burford purchased the property from the family in 1866. The following year he rebuilt the mill, adding three stories of brick to the existing limestone remains. This four story stone and brick building built in 1867 is the mill in existence today. Burford later sold the mill when he left to serve in the Missouri legislature in 1880.

After several changes of ownership the mill became part of the Cape County Milling Company in 1897. By 1914, the three mills of the Cape County Milling Company were shipping fourteen to fifteen hundred barrels of flour daily out of Jackson. Their patent flour won blue ribbons at many state fairs across the south. The Cape County Milling Company was also the first in the country to test the Alsop Electrical flour bleaching process.

Mr. Clyde Vandivort purchased the building in 1953 and used the old mill for grain storage a few years, but through the rest of the 1950s it sat idle and suffered from vandals and old age. In 1961 the Vandivort family, descendants of the Bollingers, donated the mill to the Cape County Historical Society. In 1967 the Society turned the property over to the Missouri Department of Natural Resources to become a part of the state park system. For more information call (800) 334-6946.

Story courtesy Jack Smoot.

Bollinger Mill
Photo courtesy Jack Smoot.

Dillard Mill

Missouri

Dillard Mill is a unique example of a turn of the century turbine powered roller gristmill with most of the original equipment in working order. Completion of the construction of the present mill, the second on this site, took place in 1908. The owners at that time were two Polish emigrants, Mary and Emil Mischke. In 1930 the mill sold to Lester E. Kiemme. In an attempt to keep the mill operational he added a livestock feed operation. He also opened the Old Mill Lodge where travelers could stay in the small cabins he built, fish and swim in the millpond, and eat at his table for seven dollars a day. Upon the death of his wife, Kiemme closed the lodge and locked up the mill. He protected the mill until 1974 when the property sold to L-A-D Foundation of St. Louis. The L-A-D Foundation leases the mill and property to the Missouri Department of Natural Resources, which oversees its preservation and everyday operation. In addition to the mill, there is a country store and semi-improved nature trail. For further information call (573) 244-3120.

Story and photo courtesy Jerry Wilson and Missouri Department of Natural Resources.

Edwards Mill

Missouri

Edwards Mill is an example of the milling industry traditions and the past for the students of College of the Ozarks. The mill, dedicated by Bill Cameron over twenty-one years ago, is a replica of a 1800s waterpowered gristmill. The fourteen foot overshot waterwheel, almost 150 years old, is still working. It operates two thirty inch under-runner buhrs. It also has a forty-eight inch top-runner buhr, and can operate a twenty horsepower motor for heavy production.

The timbers in Edwards Mill have supported three mills prior to this one, the first of these built in the 1700s. The late Chairman of Dixie-Portland Flour Mills, Inc., Hubert C. Edwards and Mrs. Alice S. Edwards funded Edwards Mill.

Today, the mill is operated and maintained by students and supervisors. The gristmill is located on the bottom level of the three-story building. On the level above, students weave their creations. During the Depression Eleanor Roosevelt saw weaving as an opportunity for women to supplement the family earnings. She was in-part responsible for the distribution of hand looms to various schools. For more information, contact Edwards Mill at (417) 334-6411, ext. 3355.

Story and photo courtesy Jim Davis.

Falling Springs Mill
Missouri

Mark Twain's National Forest
Orange County, Missouri
Courtesy Society for the Preservation of Old Mills

Fleetwood Mills
Missouri

Fleetwood Mills is the result of an extensive collection of "anything old". Russ Noah, owner of the mill, had several farm and home items from a farm in his family dating back to the 1800s.

Things didn't change much after the big mills cut the last of the Old Stands of timber. People just 'got by' on whatever the rough hills would provide. After WWII it became obvious that big changes were going to take place. Sawmilling modernized. At that time, Mr. Noah, after many years in forestry and conservation, decided to collect the old logging and sawmilling equipment no longer used.

A railroad tie mill, shingle mill, planer, barrel saw and tools used in these mills were installed. Also added were three stone gristmills and gas engines to operate them. The collection grew to include old tractors, Model T's and horsedrawn farm equipment.

Rustic buildings erected to house the collections were set up as a museum. Mr. Noah felt lucky to find an up-and-down mill very similar to the older ones. A six horsepower steam engine was added to power the mill. A pit saw completed a very good cross section of sawmilling until approximately 1950.

Approximately ten years ago a movement began to create a Missouri Forest Heritage Museum. Construction of the center is to begin in the year 2000. In the meantime, arrangements have been made with the museum to move most of Russ Noah's collection to the center.

Mr. Noah is seventy-six years old and glad that his efforts will be preserved as an opportunity for future generations to see how things were done in the past. For more information call (573) 226-3643.

Story and photo courtesy of Russ Noah. Photo by Wayne Sullins.

Hulston Mill
Missouri

Hulston Gristmill was established at the confluence of Sac River and Tumback Creek in Dade County Missouri in 1840.

Land records show that Ezekiel Madison Campbell, one of the original settlers of the southwest Missouri Ozarks, owned the mill site and it is assumed that he constructed the original mill.

At that time the Ozarks were still wilderness, and the mill was a focal point for a rural dispersed pioneer society. The mill served both as economical and social functions in the 1800s.

In 1848 Henry H. Pemberton bought, from his father-in-law, Ezekiel, and operated the mill until 1859. It then sold to Henery Engleman.

Engleman operated the mill another ten years following the Civil War and sold to James Christopher Hulston in 1875. In 1892 John C. Hulston, son of Cris Hulston, converted the mill to a rollermill instead of a buhr mill. He added a steam engine to provide power in times of drought. The Hulston family owned and operated the mill until 1897, when John C. Huston was murdered. In 1914 Samuel Farmer leased the mill and enlarged it to include three roller mills, with the capacity of twenty-five barrels of flour a day.

Farmer operated the mill until 1941. Other owners include Baily Floyd (1941 to 1942), Turrent Hickey, (1943 to 1946), John Pemberton Nixion and Auther Frank Nixion Jr. (1947 to 1946) and L.E. Meyer 1965 to 1967).

The operational history of Hulston Mill, spanning one hundred and twenty-seven years, came to an end on June 25, 1967 as the result of the construction of a multi-purpose reservoir on the Sac River. Hulston Mill ground its last flour June 25, 1967.

Efforts to save this landmark from destruction were sponsored by the Dade County Court, Dade County Historical Society and the Dade County Park Board. Through their efforts, the Hulston Mill was moved to a fifty-acre tract, one mile southwest of the original site. There it will be the focal point for a heritage park. Several cabins have also been moved to the site. The Hulston Mill Park is owned and operated by the Dade County Historical Society. For more information call (417) 788-2883.

Story and photo courtesy Kent Snider.

Jolly Mill
Missouri

The ante-bellum Jolly Mill, was built in 1848 with skilled slave labor. Thomas and John Isbell, father and son, began a distillery and gristmill. It sat beside spring-fed Capps Creek on the eastern edge of Newton County in Southwestern Missouri. In front of the mill ran a crude road connecting the early community of Springfield with the Kansas frontier and adjacent Indian Territory. The Isbell establishment owed the early prosperity to the prompt rise of a thriving village, appropriately called Jollification (apparently inspired by the popularity of the Isbells' product). Streams of frontier-bound settlers in wagon trains arriving at Jollification could rest, repair, and get supplies before continuing westward.

During the Civil War, all frontier traffic ceased and the Isbell establishment became idle. Bushwhackers burned the village but spared the distillery and mill.

After the war, George Isbell, nephew and cousin of the original owners, returned to operations. An effort to rebuild Jollification failed after the building of a nearby railroad bypassed the village site. George Isbell discontinued whisky making and converted the mill to meet the rising demand for flour. He later sold to his nephew, George Isbell Brown.

It then sold to the A. C. Lucas family, A. C. and his son Firman in 1895. With the mill sale, the name "Isbell's Mill" disappeared and when new flouring-rolling equipment and a modern turbine waterwheel were installed, the name became "Jolly Rolling Mills".

In 1912, the Haskins family, W.F., his son Leslie, and his grandsons, Frank and Norman acquired the mill. With the addition of a second turbine waterwheel, Jolly Mill operated at flour-making and general gristmilling for decades but eventually reverted to custom gristmilling only. Frank Haskins closed the Jolly Mill in 1973 and sold to Ben Taylor.

Jolly Mill soon became the victim of old age and Taylor began making major repairs, that included stabilizing the foundation and replacing the roof. However, a flash flood in 1978 destroyed the historic stone dam, and Taylor's restoration plans faltered.

A retired architect, provided inspiration and financial support for the local community to organize the Jolly Mill Park Foundation. One of the objectives was to restore the Jolly Mill to operating condition. In 1983, the Foundation purchased the mill and restoration started. Step by step, the mill has been renewed and demonstrates the gristmilling begun by the Isbells in 1848. Jolly Mill was placed on the National Register of Historic Places in 1983.

Story and photo courtesy Jim Moore.
Photo by Anthony Verebes.

Montauk Mill
Missouri

The last of several mills that operated in the area, this mill rests in Montauk State Park near Licking, Missouri. Montauk contains the headwaters of the Current River, which provided year-round waterpower for grist and sawmills for many decades. Montauk State Park is one of eighty parks and historic sites within the Missouri Department of Natural Resources, Division of State Parks.

Timothy Hickman had Montauk Mill designed and built in 1896, by one of the most famous millwrights in the country, William J. Furry. Some claim it took only two months to complete and considered it "an ingenuous task and outstanding accomplishment." Possibly the most unique aspect is the construction of the walls two by six feet, two by eight feet and two by ten feet boards stacked flat and spiked together. This construction practice, common for grain storage elevators, was not typical for mills. A series of millraces, gates, and a Leffel turbine were used to harness the waterpower.

Tim Hickman sold to brothers George and Alex Morton in 1903. George operated the mill and Alex ran a store until 1907, when Judge Sy Thorton purchased the property. Joe Hume, Thorton's son-in-law, operated the store, and Grover Brigman, another son-in-law, operated the mill until 1927. This included one harvest after the State of Missouri acquired the property for a state park in 1926.

Montauk Mill is representative of numerous waterpower custom mills that operated throughout the Missouri Ozarks. These mills catered to individual farmers grinding their wheat or corn into meal in exchange for a percentage of the product. Milling stimulated economic activity and frequently prompted the formation of new businesses and small communities.

The Montauk Mill is in excellent condition and one of the few remaining remnants of the old Montauk community. Most of the second floor machinery is intact and the original turbine wheel and buhrstones remain. The original roller mills (used to grind wheat into flour) were removed in 1942 and delivered to the US government for use in the World War II effort. Similar mills from the same time period have been installed for display, however, and renovation of elevator shafts and older equipment is ongoing. For more information please contact Montauk's Park Naturalist at (573) 548-2225.

Story and photo courtesy of Jamin Bray.

ROCKBRIDGE MILL

MISSOURI

In the 1800s settlers built a small community, known as Rockbridge, at the junction of Bryant and Spring Creeks. In 1841 it was named the county seat for Ozark County. The growing town was burned to the ground during the Civil War.

A large, two-story general store, blacksmith shop, bank, and church were built during the years of prosperity. During the 1930s, gasoline and diesel engines brought an end to blacksmith shops and water mills. The general store and post office were the only original buildings open by 1940.

Two brothers, Lile and Clay Amyx, purchased the buildings and the land around the mill in 1954. Lile and Edith Amyx acquired ownership of the property in 1964. A trout hatchery was built in the mill's under structure. Rockbridge has a thousand sand acres and two miles of stream.

The old mill and buildings are now an office, restaurant, and lounge. Rockbridge Mill, now known as the Rainbow Trout Ranch, is tucked away in a picturesque secluded valley and its beauty lives on.

Story and photo courtesy Sandra Amyx. Photo by Gene Kletchka.

Watkins Woolen Mill

Missouri

Watkins Mill is the only remaining nineteenth century woolen mill in North America with the original machinery. Waltus Lockeff Watkins, a successful livestock dealer and farmer, constructed the mill in 1860. Watkins, who had worked in textile mills as a young man, had previously added brick kilns, a gristmill, and a sawmill to his 3600 acre farm in northwestern Clay County, Missouri.

Although mid-sized, compared to eastern woolen factories, in the mid-west it was a large mill, and three times the size of most of Missouri's textile mills.

The three-and-a-half floor brick factory produced a dozen kinds of fabrics, shawls, several kinds of blankets, knitting yarns and batting. It operated from 1860 to1898.

The mill furnishings included Jenks & Sons and Furbush & Gage carding, spinning, yarn processing, weaving and finishing machines from Pennsylvania. The sixty horsepower steam engine, manufactured in St. Louis, was supplied by a large steamboat boiler. Opening on the eve of the Civil War, the mill prospered as the demand for cloth grew throughout Missouri and eastern Kansas. Additional carding machines and looms were added in 1863. The business continued to grow. More looms and an additional spinning machine were added in 1868. Sometime in the 1860s the gristmill was moved into the factory building to take advantage of the steam engine. Watkins employed some forty skilled mill operatives, a number of whom were European immigrants, as well as, a miller and two helpers.

In the 1880s the woolen industry began to fail in the midwest and many mills closed. Finishing machinery failure, in 1886, caused Watkins Mill to cease fabric production. In 1892 they stopped making yarn and did only a small amount of custom carding and occasional blanket production. In 1898 the factory was closed completely, although the gristmill ran until 1905 and the farm continued to operate well into the 1930s.

In 1870 there were over 2400 woolen mills in the United States. Watkins Mill is the only one remaining with its original equipment still in place. In 1966 it was designated a National Historic Landmark. It was named a National Mechanical Engineering Historic Landmark in 1980.

Watkins Mill is operated by the Missouri Department of Natural Resources, Division of State Parks. For more information contact Watkins Woolen Mill State Historic Site & Park at (816) 580-3387.

Story and photo courtesy of Missouri Department of Natural Resources.

Wommack Mill
Missouri

Constructed in 1883 by John Boegel and Joseph W. Hine, the gristmill at Fair Grove became northern Greene County's most prosperous business. Various owners offered services to local farmers continuously until 1969. A stationary steam engine belted to a series of shafts and pulleys transferred power for shelling and cleaning processes. Corn was reduced to edible meal by a pair of forty-two inch buhrstones. Nearly all mills in the Ozarks, including several in the Fair Grove area, were waterpowered until after the Civil War. This served as a free source of energy but constantly threatened mills and adjacent dams with destructive floods or created problems due to the lack of flowing water during drought. Progressive businessmen, Boegel and Hine could operate their steam mill any day of the year.

The original single "run" of horizontal grinding stones, still in place, are French buhrs. Sometime during the first decade of the twentieth century a more modern system of milling was introduced. A set of three steel roller mills, belted to a long line shaft in the basement, efficiently converted wheat into baking flour. Although those roller mills are long gone, the receiving separator, double scouring machine, flour dresser, and dust collector from that era remain.

A 1935 inventory of equipment does not list the steel roller mills, but has the forty-two inch buhrstones. There was probably little or no resale value for these "out of date" items. T. Frank King was operating the mill at this time, and it is supposed that his business direction changed from grinding meal and flour to producing only livestock feed. He introduced an industrial hammer mill for this purpose, powered by an electric motor and later a gasoline engine.

The Wommacks purchased the mill in the early 1940s and brought it up to date with a feed mixing machine. About 1950, he installed a diesel engine with plenty of power and lots of noise, according to the residents of the previously sleepy little town. The old mill ceased its productive existence in 1969, upon the death of Clifford Wommack.

Once the scene of business and social activity, the forgotten building stood vacant and deteriorating until 1984 when the Fair Grove Historical and Preservation Society purchased it and two acres of land from the Wommack family. Since that time, members have been able to get the Wommack Mill listed with the Greene County Historic Sites Board and as a National Historic Site. A major restoration effort began and continues still. For more information call (417) 759-2742.

Story and photo courtesy of Dan R. Manning

Champion Mill

Nebraska

Champion Mill is the last functional waterpowered mill in Nebraska. In the 1880s southwestern Nebraska began evolving from the ranch era into the agricultural settlement period.

In 1888, Thomas Scott completed construction of a mill in the town of Hamilton on the Frenchman River in Chase County. When bidding for the county seat in 1887, residents changed the name to Champion. Imperial won the county seat but the town's name remained Champion.

After receiving his waterpower rights in 1887, Scott started flour and grist milling at the Champion Roller Mills company. According to the September 5, 1889, issue of the Chase County Champion newspaper, the mill was equipped with "... machinery of all the latest patterns. First class work guaranteed." The mill did a brisk business in flour and other mill products. In the early 1890s fire destroyed the mill and the ice skates and coats of local children who used the millpond for winter recreation. By 1893, R. P. "Dick" James had rebuilt the mill and operations resumed.

The mill was a gathering place even in 1897, when the road crossed the dam and overflow.

Little is known of the 1888 mill, but the 1893 mill was a two-story, twenty five foot by thirty-two foot rectangular building. It had a stone foundation and housed more sophisticated machinery than earlier mills in Nebraska.

Two new processes in flour milling were being used by American millers in the 1870s and 1880s. Scott used steel rollers instead of the old buhrstones commonly used in milling until the invention of the gradual reduction roller process in the late 1870s. Using the new roller system coupled with a new process of sifting and regrinding undoubtedly gave Champion Mill an edge in an evolving industry by offering a better quality of flour.

For more information call (380) 394-5118.

Photo and story courtesy of Michael E. Hoskovec.
Color photo by Dennis Spady.

Champion Mill has long been a recreation center, as 1921 photo attests. Then, swimmers could rent suits at a bathouse for 10 cents.

The Florence Mill

Nebraska

The Florence Mill sits at the crossroads of Nebraska's pioneer trails near the Missouri River. Brigham Young supervised the building of the original mill in 1846-1847 during the Mormon Trail era. The 1849 California Gold Rush and alternate Oregon Trails both passed by the mill site during America's nineteenth-century western migration.

The Florence Mill contains the only structural remains from the Mormon settlement of Winter Quarters in Indian Territory. Winter Quarters and its gristmill were abandoned in 1848 as the land and improvements reverted back to the Omaha Indians.

The gristmill was an abandoned ruin in 1849. Alexander Hunter revived the site around 1856. Hunter was a gold rusher who decided to settle in Nebraska Territory. The mill contributed to the development of Nebraska Territory after it was opened to non-Indian settlers.

Jacob Weber, a baker and a German emigrant, worked at the mill after his bakery failed in the Panic of 1857. Within three years he became known as the miller of Florence. The Weber family owned and operated the mill for three generations, from 1860 to 1964. The Weber Mill became a landmark and reminder of Florence's origins.

The Florence Mill remains an example of industrial Americana. Douglas County once had twenty mills, but only the Florence Mill and two others remain. The old mill stands today because it evolved to fit changing times. The mill changed from a waterpowered grist and sawmill to a roller mill system for flour and conversion to a steam-powered mill in 1888. The addition of an elevator took place when conversion from a flourmill to a feed mill was done. In 1923, electricity replaced steam power.

The Depression years were hard on the mill. The mill sustained damage in a 1937 hailstorm and flooded to a depth of fourteen feet. Two years later the mill was moved to higher ground away from Mill Creek. Wooden pegs such as those used in the construction of the Mormon Tabernacle in Salt Lake City were discovered during the move. The mill closed from 1943 to 1945 while its owner, Lyman Weber, served in the armed forces during World War II. Twenty years later he sold the family mill due to health problems.

Progress threatened the mill in the 1970s with construction of the Interstate Highway but owners Ruth and Earnie Harpster protected it from demolition. Grain continued to be processed at the mill for the next decade, until once again abandoned in 1989.

Though remodeled, moved and added to, the Florence Mill still uses original hand-hewn and pegged beams to support the second floor. Marks of the Mormon workmen's chisels still show. The mill continues to use the original rope-driven grain elevator and cleaner added in 1915. The integrity of the mill has not changed in over 150 years. The Florence Mill has become a monument to the pioneer spirit, standing as a silent witness to the past.

The Florence Mill was listed as the Weber Mill on the National Register of Historic Places in 1998. Restoration is in its early stages.

An artist, Linda Meigs, purchased the dilapidated and vacant mill in 1998. The dirty work of cleaning-up a decade of neglect has been started.

The Florence Mill will become a Winter Quarters Mill Museum, art gallery and educational site for students of history. The museum will interpret pioneer mining and the evolution of milling as an industry, and memorialize the profound ideas embodied by the historic Mormon Trail. For more information please call (402) 551-1223.

Story and photo courtesy Linda Meigs.

Neligh Mill

Nebraska

The Neligh Mill State Historic Site is operated by the Nebraska State Historical Society. The mill, built in 1873, is the only nineteenth century flour mill in the State of Nebraska with all of the original equipment intact. The mill complex includes the main mill building, elevators, office, reconstructed flume and penstock. The history of the Neligh Mill spans almost one hundred years, from construction in 1873 to its final days as a feed mill in 1969.

Construction of the brick mill building began in 1873 by the town's founder, John D. Neligh. W.C. Gallaway, an early Neligh businessman, assumed ownership and saw to the completion of the structure. Gallaway improved the dam on the Elkhorn River, rebuilt the flume, and installed a horizontal water turbine to power the drive works in the basement of the mill. Two runs of stone buhrs were put in, and the mill commenced grinding on August 29, 1874. In 1878 two more stone buhrs were added.

In 1886 the stone buhrs were removed and replaced with seven stands of Allis rollers. The new rollers increased capacity to 140 barrels of flour a day. Along with commercial orders, government contracts were received from the Indian Bureau and War Department. The mill operated twenty-four hours a day as much as possible.

Around the turn of the century, the mill increased its capacity to 250 barrels a day and eventually 500 barrels a day. From 1900 to 1920 Neligh Mills was one of the top producing flour mills in the state. To more efficiently utilize the river waterpower, the dam was rebuilt and the drive cable from the penstock was replaced with a twelve inch wide leather belt. Hydroelectric power for the city of Neligh was produced with a horizontal turbine from 1900 to 1915. The growing demand for electricity in 1915 brought the installation of a 225 horsepower diesel engine to provide power for the generator until 1925.

The Depression of the 1930s forced many Nebraska mills to close, but Neligh Mill remained open. World War II brought an increase of flour production, but after the war production dropped. By 1959 the mill ceased to manufacture flour altogether, but continued to grind livestock feed until 1969.

Exhibits on the mill history and operation are now found in the 1915 addition that once housed the electrical power plant. For more information contact the Neligh Mill State Historic Site at (402) 887-4303.

Story and photo courtesy of Don Ofe.
Postcard courtesy of
Society for the Preservation of Old Mills

Windmill State Recreation Area

Nebraska

Windmill State Recreation Area in Gibbon, Nebraska is one of two highly developed areas operated by the Nebraska Game and Parks Commission in the state's unique "chain of lakes."

It draws its name from the locale, Windmill Crossing, where the Pawnee Indians forded the Platte River during their annual buffalo hunts. The old-time windmills that add so much to the character of the area were acquired in the early 1970s. Collected from around the Midwest, they have been restored to working order.

The largest is a railroad windmill that stands over sixty feet tall. The wheel is twenty feet in diameter, with 200 blades in two rows of eight sections each. The vane or small fan section that extends beyond the edge of the wheel is actually a type of governor. When the wind is too strong, it turns the head out of the wind and shuts it down so it won't spin at excessive speeds that can damage the mechanism. The main tower extends forty feet to the platform, and the "stub" tower rises another eleven feet. First erected in a small town in northeastern Colorado, probably Fleming, about 1890, it is the oldest windmill at the park. Taken down in 1955 and moved to Kansas, it was found at "Sodtown," just east of Colby, and moved to Garden City. After acquisition on August 1972 from Homer Sargent of York, Nebraska, the relocation to Windmill State Recreation Area took place. After reconstruction, it was erected here on November 1, 1973.

The smaller Waupans windmill, first erected in 1902, came from Louis Drake of Arcadia, Nebraska, and was placed here in the fall of 1972 after it its restoration. The ten foot, folding wheel always faces downwind, while the weighted tail points into the wind.

Built about 1910, the Dempster Model 9 windmill came from southwestern Nebraska. Its ten-foot, flat wheel faces into the wind. Dismantled by Paul ten Bensel of Wilsonville and Willis Bentley of Cambridge, it was renovated before being erected here in July 1973.

Story and photo courtesy of Mark Clymer, Windmill State Recreation Area.

Standard "Railroad" Windmill

Dempster Model 9 Windmill

Waupans Windmill

Belknap Mill

New Hampshire

According to preservationists at the Smithsonian and Boston University, the Belknap Mill is the oldest unaltered brick textile mill in the United States, and the only remaining one that represents the first stage of the Industrial Revolution in America.

Built in 1823, along the Winnipesaukee River in Laconia, New Hampshire, it was among the first mills in the country to convert from weaving to knitting. This took place about the time of the Civil War. It remained a hosiery mill until 1969.

In the early 1970s the mill was converted into an art and history center. It now houses the nation's first permanent exhibit on industrial knitting.

The Belknap Mill Society was the first organization in the country to receive federal funding, and a national award from the National Trust for Historic Preservation, for preserving an industrial structure.

In 1998 the Belknap Mill was one of three museums to receive the National Award for Museum Service for downtown revitalization and service to the community. For more information call (603) 524-8813.

Story and photo courtesy of Mary Boswell.

Al Daloz Mill

New Hampshire

The Al Daloz mill is a small wood processing mill on the lower reaches of Ferguson Brook on the south end of Hancock, New Hampshire. The present millsite dates around 1797, when Capt. Bradford and his family settled and erected a saw and gristmill several hundred feet downstream of Hancock's earliest mill built in 1783 by Samual Ames.

Above the mill are large wetlands that moderate the streamflow to an average of about ten cfs for most of the year. This is said to be considerably less flow than when the mill operated in the 1800s. The hillsides upstream were primarily pasture and orchards at that time. The two acre millpond is formed on the south side by a small glacial moraine, with a ten foot tall wooden-faced stone dam. The mill sits along the east side on the last significant drop in elevation before the wetlands.

This mill, typical of many of the small mills erected in the 1800s on the streams of New England, used waterpower for local needs and manufacturing. The "Daloz" mill served various owners and purposes through the years, processing lumber and manufacturing boxes, barrels, shingles, wheelbarrows, and other wooden items of value in their time. Danforth and French, the owners at the time, removed the flour milling equipment in the 1860s.

By 1900, boxes and barrels from the mill were used for shipping apples to Boston and England. Ice was harvested from the millpond during winter, slipped over the dam into an icehouse and stored in sawdust for sale in the summers.

The mill has been known as Capt. Bradford's Mill, French and Danforth Mill, Old Johnson Mill and now the Al Daloz Mill. It is the only remaining mill in town of at least sixteen active in the past. Few, including this one, figured prominently in the New Hampshire state industrial censuses.

It appears to have been a part-time mill, operated at the whim of local needs and weather, although it certainly has a long history in the town and the area. Fred O. Johnson operated the mill through most of the 1900s to support his wife and seven children. His son, Willis Johnson, continued to work the mill into the 1970s, although last operated commercially in 1961.

Al Daloz purchased the mill in 1958. Aided by his wife, Lois, development of the mill continued. A decayed five foot diameter turbine and the log saw equipment were removed, the dam and the Francis turbine refitted, various antique woodworking machinery and tools added, and a second sluice in the dam for water rams and a small waterwheel reopened. Daloz had an enthusiasm for belts, pulleys, waterpower, antiques and innovations. He polished stones, planed lumber, built birdhouses, boxes and planters but never turned much of a profit. Al Daloz died in 1992.

Power, currently provided by a thirty-four inch Francis Turbine runs planers, tongue and groove, edging, cordwood, and bolter saws. A rebuilt nine foot overshot waterwheel powers wood-turning lathes and water pumps for nearby gardens. Operations can be supplemented in winter by a Caterpillar tractor engine housed in an attached shed. The property is maintained by Lois and his son, Charles. It is operated sporadically as a full-of-hopes-and-potential organic vegetable garden and wood processing mill.

Story and photo courtesy of Charles Daloz.

Littleton Gristmill

New Hampshire

In 1797 a resident of Littleton by the name of Solomon Mann purchased the mill privilege on the Ammonoosuc River. There he built a dam, a saw mill and gristmill. The state-of-the-art gristmill had many of the features described in Oliver Evans Handbook. The prime source of power, an overshot wheel, turned millstones and grain elevators and powered sifters. The three and one half story mill building had clearstory windows to divide the roof lines. The foundation wall formed the bank of the river. By the turn of the century, the riverfront mill area had become the nucleus of a prosperous settlement. A tannery, store and fulling mill were also added.

After several years the gristmill sold to millwright, Asa Lewis, who operated it until his death in 1815. Thereafter, it changed ownership frequently. Circa 1890 conversion to the use of Hercules water turbines, which powered a Robinson mill began. It operated with varying degrees of commercial success until the 1930s. It was then used by a door and sash manufacturer and storage, and allowed to fall into a state of profound disrepair.

In 1997 two local families formed Renaissance Mills of Littleton and undertook restoration of the gristmill and three other mill buildings, thought to be the oldest commercial buildings north of Concord, New Hampshire. Goals included restoring the basic workings of the gristmill as a historically accurate replica of the original and open the mill as a working museum. Plans also include restoration of the other mill buildings as commercially viable retail and rental space, thus creating an eigthteenth century riverfront complex.

Today the look and feel of the original construction has been maintained wherever possible. The exterior is clad in radial sawn spruce. The roof is hand-split cedar shakes, wide pine flooring on the three floors is fastened with wrought iron nails, twelve over twelve true divided light windows mimic the originals, as do the clearstory windows on the third floor. All hardware and the interior and exterior lanterns are hand forged. The fifty-five foot chimney is replicated in used brick, and the interior walls are covered with a plaster skim coat.

The same careful research application was used for replication of the working mill. The exterior undershot water wheel measures twenty feet six inches in diameter. Current regulations prohibit the construction of a dam necessary to power an overshot wheel. The wheel drives interior wooden gearing in the basement, which in turn drives a four foot diameter gristmill stone on the first floor of the building. The wooden gearing is made of indigenous woods typically used 200 years ago. The metal hubs, bearings and shafting recovered from the mud in the mill basement, guided the design of the gears and drives. Other components were designed and made according to state-of-the-art practices used 200 years ago. The wooden gearing and attached shafts are housed in a hearst frame made with timbers similar to some of the frame components found in the mill. With the exception of metal shafting, metal components and hardware, all design and construction occurred in-house.

Other artifacts recovered provided invaluable information as to the size of the original millstones, which were four foot in diameter. A matched pair found ten miles upriver could have been the stones grinding grain in 1798.

Littleton Gristmill was restored by private funds and is now a working museum. The mill also houses an authentic nineteenth century gristmill which uses twenty-four inch stones and is operational today. For more information call (888) 284-7478.

Story and photo courtesy Ron Murro

Scheyd-Yeaton-Kayes Mill

New Hampshire

In 1973 Harold F. Yeaton purchased property in Washington, New Hampshire, containing a house built in 1805, a barn, and the remains of the town's first mill, built in the 1770s. The town was the first in America named after President George Washington, in 1776. Over the years, the mill sawed lumber with an up-and-down saw, made apple cider, and sold ox yoke beams. Later, the mill made tubs and shingles, and operated for a time as a gristmill.

In 1977, with the intention of perhaps reconstructing a mill on this site again, the Yeatons purchased twenty-three beams measuring twenty-two feet by fourteen feet by twelve inches. These beams were salvaged from the Amoskeag Mills in Manchester, New Hampshire. Miscellaneous equipment was collected over the years. In 1978 the decision had definitely been made to carry through with the construction of a new mill.

A deal with a demolition contractor provided for the salvage of materials from a building with a hand-hewed timber frame and 200 year old side boards. Long-time friend, George Kayes, donated $1,000 to buy the materials. Engineer/Contractor and friend, Louis Scheyd, agreed to lay out the floor plans and do the excavation for the footings. The materials were piled up in Scheyd's construction yard and a roof was built over them to protect them from the weather while building the foundation.

The new foundation placement would preserve the original turbine pit and the underground culvert that conducted the tailrace water back to the river. It took several years to construct the foundation.

Harold Yeaton built a Ford Power Unit in 1952, while working in a Ford dealership where they built new units for sawmills. This was the first machine moved into the mill followed by the ironwork for reassembling the up and down saw.

A twenty-four inch planer, twelve inch jointer, thirty-six inch bandsaw, table saw and chamfer machine were acquired. Three machines built by Yeaton's father in the 1930s, an air compressor, grinder for sharpening axes and grindstone for sharpening scythes were added. Next came the installation of an eleven foot by fifteen foot Erie steam engine. There is other machinery in the mill, including a Lane shingle machine. The last few years have been spent setting up the machinery.

Story and photo courtesy of Harold Yeaton.

Cooper Mill
New Jersey

The Cooper Mill was built in 1826 in Chester, New Jersey, as a testament to the power of water to grind grain. The mill uses water from the Black River, which meanders through flatlands until it reaches the quaint village of Milltown, Chester Township. Here the land drops and the Black River begins to flow rapidly. Several of the original Milltown homes, including the historic Abram Cooper house, have survived.

Prior to 1761, Isaiah Younglove realized the importance of the area for waterpower. He built a gristmill at the start of the rapids. Isaiah listed the mill and 160 acres for sale in 1761.

Thomas Fairclo owned the mill until about 1788, when Elias Howell bought the property. In 1825, Nathan Cooper, a bachelor, purchased the gristmill, the millpond, a sawmill, and four and one half acres, for $750 to settle the estate of Elias Howell. The following year, he built the present mill.

The new mill on the Black River, large for its era, had four runs of millstones and two waterwheels. The design followed the modern milling techniques of Oliver Evans, which included elevators, augers and other machines to mechanically move the grain and flour throughout the mill, saving manual labor.

The stone mill has four levels. The miller's office and four runs of millstones are on the main floor. Cooper proved to be contemporary with his choice of French buhr millstones, preferred for the end product of whiter flour. From the millstones flour dropped to the lower floor and into elevators to be carried to the upper floors and bolted or sifted. The lower floor also contains the main gearing to transfer the power from the waterwheel to the millstones and machinery on various floors of the mill. On two upper floors are the grain cleaners, bolters, and grain storage areas. Flour bolters use fine cloth screens to sift the flour into several grades.

Nathan A. Cooper inherited the mill from his Uncle in 1826. The Coopers hired millers to run the gristmill for them. Nathan owned the mill until his death in 1879, when his son Abram inherited it. Abram leased out the mill and eventually sold about 1913.

In 1963 the Morris County Park Commission obtained the mill and began restoration. In October 1978 the mill opened, operating one run of millstones. The mill is still powered by water from the Black River that turns a Fitz overshot waterwheel. Two of the four runs of millstones are operational and one or both are used to demonstrate milling techniques. For more information call (908) 879-5463. Courtesy Morris County Park Commission.

Story courtesy Ivins Smith, Miller/Historian.
Photo by Don Weaks.

HISTORIC SPEEDWELL MILL

NEW JERSEY

Historic Speedwell in Morristown, New Jersey was primarily an industrial complex. Situated at the base of a gentle slope, water for the overshot wheel came from a small pond uphill through a wooden flume. The flume is concealed in the rafters of the wheelhouse. As the water flows away from the base of the wheel through a spillway it gives the impression that the wheel is turning against the flow of the water.

The original structure, a one story building with a garret and cellar dates back to the late eighteenth century. In 1830 Stephen Vail, owner of the Speedwell Iron Works, purchased the factory from his son-in-law and added a second story to house cotton weaving looms.

During the winter (1837-1838), Stephen lent the vacant second floor to his son Alfred and Professor Samuel F. B. Morse to be used as a secret workshop. It was here that the electromagnetic telegraph was perfected, and the dot and dash alphabet written by Alfred and his apprentice boy, William Baxter. The telegraph was first publicly demonstrated to 345 locals on January 10, 1839. Designation as a National Historic Landmark occurred in 1975, due to the historical significance of the event. A grant from the National Endowment for the Humanities created the telegraph exhibit on the second floor.

In 1850, a gristmill was installed in the first floor and basement, a wheelhouse constructed and the smaller waterwheel replaced with a twenty-four foot overshot wheel, the largest operating wheel in New Jersey. The wooden flume was replaced with a twelve-inch underground pipe. A "Bone Grinder" was installed in the wheelhouse to make fertilizer and shafts were installed to operate the mills and equipment. The mill on occasion was rented to locals or in barter for services provided to Speedwell.

In 1968 Historic Speedwell acquired the property. The factory had deteriorated and the building had to be shored up to prevent collapse. Basement walls were replaced and put on concrete footings. The basement stairs and a portion of the first floor had to be replaced. The roof was replaced with new wood shingles. Tracks were laid to roll the waterwheel out for restoration. A fresh coat of paint, in the original color, returned the building to the 1838 appearance. For more information call (978) 540-0211.

Story by Harold Benz.
Photo courtesy of Kathleen Q. Duane.

Kirby's Mill
New Jersey

Originally Kirby's Mill consisted of a complex of six separate buildings, the gristmill (four stories and basement), storage barn, two sawmills (one rotary and one up and down), combination cider-shingle mill, and combination wagon/blacksmith shop.

The current mill consists of the gristmill, storage barn, a completely restored and fully operational blacksmith shop, and a one-third scale model fully operational rotary sawmill.

The gristmill restoration is ninety-five percent complete. When finished there will be two fully operational grinding platforms. The waterwheel and gearing are now fully operational and demonstrated at events at the mill.

The miller's house across from the millrace is fully restored and occupied.

The first white settlers came to Medford in the late seventeenth century and were members of the Religious Society of Friends, a peace loving group who had nothing but the friendliest relations with the Indians in the area. Although the last remnants of the Indian population have been gone from Medford and the state for over 150 years, the area is immensely rich in Indian heritage, particularly in Old Eveshan Township (which included Medford until 1847). Medford became a separate township in 1847. For more information call (609) 654-7767.

Story and top photo courtesy Thomas N. Chavis.

Clydesdale and Jumbo Trucks used for transporting goods. Charles Kirby seated in truck on right. Gristmill partially shown on left and cider and shingleing mill or right (1919). Photo courtesy Society for the Preservation of Old Mills.

Moravian Mill
New Jersey

The gristmill in Hope, a four-story structure, was said to be the tallest building in New Jersey at the time of construction. Limestone was used and quarried about a mile away. The project required a 150-yard long millrace, partly cut twenty feet deep through shale. The mill featured an internal overshot wooden wheel replaced around 1900 by a Fitz wheel, that functioned until the early 1950s. The last years of the mill usage declined when most of the area dairy farms shut down.

During the Revolutionary War the mill supplied flour to General Washington's army during their winter encampment at Jockey Hollow (Morristown). The designer and supervisor of the construction, Hans Christian Christiansen, was so highly regarded for its technical excellence that General Lafayette sent an aide to prepare a report about it.

After the Moravians sold Hope community in 1808 a succession of millers, about which little is known, had ownership.

In 1985 the vacant mill underwent conversion into the seventeen room Inn at Millrace Pond. The skeleton of the wheel is a prominent feature of the restaurant and tavern. Other parts of the mill have also been preserved. Portable equipment was given to the Moravian Museum in Bethlehem in the 1970s. For more information call (908) 459-4884.

Story and photo courtesy of Charles Puttkammer.

Waln's Mill

New Jersey

The bright blue gristmill at Historic Walnford is a reminder of the changes in taste and technology that have occurred over the last century. Originally built in 1735, the gristmill became one of the earliest structures in what would become the village of Walnford. On the banks of Crosswicks Creek in Upper Freehold, New Jersey a fulling mill and sawmill were added to this early industrial site along with the supporting trades of the cooper and the blacksmith. By the end of the eighteenth century, Walnford had become a thriving production base for an international trade route.

Richard Waln recognized this potential when he purchased the property in 1772 and named it Walnford. He added a third set of French buhrstones in the gristmill and built a Georgian home, thought to be the largest in the county at the time. According to his ledgers, line white flour, lumber, wool cloth, apples, peaches, barrels of lard, and smoked hams were loaded onto boats that traveled downstream to the Delaware River and the port of Philadelphia. At Waln's wharf goods were transferred to sailing ships and sent to the Carolinas, Barbados, Spain, Portugal and England.

Nicholas Waln inherited the property from his father when the gristmill burned in 1822. He took that opportunity to enlarge the mill. Today his initials and that date, carved into a stone of the foundation, can still be seen. As the country grew, less of the nation's grain was grown in New Jersey. The mill shifted from export production to serving the local community. Sadly, a fire again destroyed the mill in 1872. The third generation at Walnford, represented by Sarah Waln Hendrickson, went against the advice of friends and family and rebuilt the mill. Using her father's foundation, a line timber frame structure was rebuilt and stands today. Powered by a four foot diameter Risdon turbine, the gristmill is an example of a late nineteenth century automated stone mill.

The mill continued to operate as a custom mill for flour and feed until 1917, when the owner, Richard Waln Meirs, Sarah's great-nephew, died. The mill closed and remained almost untouched until the mill and the surviving buildings at Walnford were given to the Monmouth County Park System by the Mullen family.

The Park System has undertaken the restoration of the mill and milling equipment. The vivid blue paint on the exterior of the mill is the result of a paint analysis dating several layers and colors. Unfortunately environmental changes have altered the flow of the creek so the mill no longer has the necessary head at the dam to power the turbine. It now has an electric motor that drives a hydraulic system. The mill is operational and has a variety of exhibits describing the Waln Family History. In addition to the mill, the house, carriage house and a number of farm buildings have been restored to their late nineteenth or early twentieth century appearance. For more information call (609) 259-6275.

Story courtesy of Sarah Bent.
Photo by Bradley Olman.

Carmon Phillips Old Mill

New Mexico

If these Old Mill walls could talk, the stories they could tell of Billy the Kid, Pat Garrett, Geronimo, and "Black Jack" Pershing. They inhabit the dark places of the old mill where Delana Clements grew up with echoes of what was once the wild, wild, West. Her parents, Carmon & Leona Mae Phillips, bought and restored the historic Old Dowlin Mill. It was the first major building constructed by white men at the foot of a sacred Apache mountain in Lincoln County, New Mexico. It was built at the end of an era, the last vestige of that untamed territory and the beginning of Ruidoso.

Long before the white man came, the mountain wilderness of Sierra Blanca had been home to various Pueblo and Nomadic Indians and later to the Mescalero Apaches. The land belonged to them.

In 1866 in this alpine setting Captain Paul Dowlin decided to settle and build his first water mill. The mill was promptly washed away in a flood. He called for his younger brother, Will, to come from the East. Together they salvaged the machinery and built the current mill on higher ground in 1868. Water came down a trough by a series of diversion ditches, and over a double-wide overshot waterwheel. It powered a gristmill, sawmill, bean thrasher, and blacksmith shop. It served as a trading post, dance hall, and Ruidoso's first Post Office. In 1878 the famed Lincoln County War broke out between competing businessmen. One of the businessmen, John Tunstall, hired a young boy named Billy the Kid. Billy was good natured and well liked by many, including the Dowlin brothers. He spoke Spanish and charmed the local girls at the dances. After witnessing the murder of his boss, Billy sought revenge. He was later charged for murder and pursued by Sheriff Pat Garrett. By this time, the mill belonged to the Lesnett family, and it is rumored that they hid Billy from the law in a flour barrel.

There were several owners after that, including the Charlie Wingfield family whose children were born in the mill. Wingfield's son, Ike, moved the post office to another location and the village name changed from Dowlin's Mill to Ruidoso, which means "noisy water".

At the turn of the Century, the mill was abandoned for many years, but Ruidoso grew into a thriving mountain resort. In the late forties, the Phillips family restored the mill to working order and opened a gift shop and photography studio. They have owned Carmon Phillips Old Mill (Dowlin's Mill) for over fifty years and still use the waterwheel in the summer to demonstrate the milling process. For more information call (505) 257-2811.

Story and photos courtesy of Delana Clements.

Cleveland Roller Mill

New Mexico

The Cleveland Roller Mill, also known as Cassidy's Mill and built by Joseph J. Fuss, was the last mill built in the valley. The majority of the mill equipment installed in 1899, production began in 1901 and ceased operations in 1947.

The mill building is primarily adobe and has changed very little. It used rolls for grinding wheat and had the most modern sifting and separating equipment available at the time. The west half served as the granary and the east half contained the mill machinery. The waterwheel sat on the west side of the building. The mill could process about fifty barrels (166 bushels) of flour or corn meal in a twenty-four hour period.

It is called a roller mill because of the cold steel rolls used to process the wheat. This technology imported from Europe in the 1870s, began to be used in American mills by the 1890s. The mill also had a semi-automatic transport system. Prior to the use of this system, it would take as many as seven men to perform the milling operation. With the new system it took one to three men. Belt mounted buckets, screw conveyors and wooden spouts moved the grain from one bin or machine to another. Also used were various cleaners, separators, sifters, purifiers, flour and bran packers. An eighteen foot six inch waterwheel, purchased and then assembled on location, provided the power.

Two long main shafts drive the nineteen machines, nineteen elevators and seven screw conveyors. There are forty-two belts, ranging in width from one to fourteen inches and in length from one to eighty feet with one hundred feet of steel-link drive chain. The belts were made of leather, laced together and held tight and adjusted with a tightening block. If the belt stretched too far it was cut and replaced. Later belts were made of rubber or canvas.

In cases of water shortage, a wood and coal fired steam engine outside the south wall was used. The first steam engine exploded in 1921 killing Joseph Fuss's nephew. Although the engine and building housing were rebuilt, both were dismantled and sold in 1957.

The Cleveland Mill received the first contract to provide electrical power to the community in 1913. Daniel Cassidy, Sr., an Irish immigrant, purchased the mill and later sold it to his eldest son. It operated until 1947 when the decline of wheat farming and population made it unprofitable to run.

In 1979 Cassidy's grandson procured placement of the mill on the state and national historical registers. In 1984, with the help of grants, Cassidy began the first of five stages of restoration that resulted in the mill opening as a museum in 1988. This is the only mill in the area that kept the original machinery intact.

The history of grain milling in northern New Mexico began around 5,000 BC with the pre-historic Anasazi and continued through the Spanish conquest and settlement in 1598, to the Anglo influence in the 1840s, and mid 1900s. The Historic Mora Valley Foundation at the Cleveland Roller Mill Museum has preserved much of this milling history and technology. For more information call (505) 387-2645.

Story courtesy of the Society for the Preservation of Old Mills.
Photo courtesy of Daniel Cassidy.

El Molino Grande De Sapello
New Mexico

El Molino Grande de Sapello is a replica of the original mill erected by the Pacheco family in Sapello, New Mexico. This mill was built in the early 1800s and located north of Las Vegas. Because the Sapello millhouse was constructed with adobe bricks, it could not be moved. The original milling equipment, manufactured in Buffalo, New York and shipped from the east coast to New Mexico by railroad, still operates in the replica. The mill ground wheat for both the U.S. Army and the civilian population until the 1930s when chronic crop failure and changing economic conditions combined to force closure. Transportation improvements made it convenient and less expensive for people to drive from the Sapello area to Las Vegas, New Mexico, and buy commercial flour rather than having their grain milled. At that time the Leger family, who acquired the mill in the early 1900s, were the owners. The equipment underwent various upgrades over the years.

In 1968 Y.A. Paloheimo, the museum founder, acquired the mill for El Rancho de las Golondrinas museum. All of the milling equipment and metal components for the wheel were removed from the original structure in the replica. At that time a new pine waterwheel was installed. Since 1968 some worn gears have been replaced with newly cast gears, and in 1990 the waterwheel was rebuilt using thin plate steel buckets and white oak from Arkansas. It is hoped that the museum now has a mill that will last for generations.

The floor on which the millstone operates is separated from the rest of the flooring in order to minimize the vibrations and preserve the structure of the adobe building.

El Molino Grande is one of four mills at El Rancho de las Golondrinas, a living history museum. Three mills are small with horizontal waterwheels and only one is operational. Las Golondrinas was a *paraje,* or overnight stop, on El Camino Real, the Royal Road, to Mexico City. "The Ranch of the Swallows" was founded in the early 1700s and has been a living history museum since 1972, interpreting the eighteenth and nineteenth centuries. On its 200 acres are more than thirty original, reconstructed and related buildings brought in from other sites. For more information call (505) 471-2261.

Story and photo courtesy of Louann C. Jordan.
Photo by Doug Peterson.

Valencia Flour Mill

New Mexico

Flour milling came to Jarales in 1914 when Jose Dolores Cordova I built a small structure where the current tin and wood frame mill stands. Wheat was a staple crop grown in the Hispanic village of Jarales from the late 1880s to the 1930s. Winter wheat, especially abundant, prompted Cordova to establish a flour mill in the center of Jarales. The original mill, Jarales Trading Co., Inc., was incorporated on Dec.17, 1913. With his partner, Jose Garcia, a businessman in the nearby town of Belen, Cordova sought to improve the area's economic base and profit from local wheat shipped to Socorro and Albuquerque. Cordova charged a fee of fifty percent of the wheat local farmers brought for grinding. By 1919, the partners changed the name to Jarales Roller Mill. Steel rollers driven by a diesel engine connected to a line shaft were used. Heavy cedar beams from the original mill can be seen in the whitewashed basement today. Unfortunately, the rest of the first mill burned in the early 1930s.

A few years later, Cordova's son, Arturo Cordova, Sr., became partners with his brother-in-law, Vivian Sanchez and reopened the Jarales Roller Mill. After Sanchez died in the early 1940s, Arturo continued running the "family flour mill" until the 1970s. When Arturo died in July 1983, the mill fell into disrepair until Jose and Kathryn bought the mill properties from the Cordova family estate in August 1988.

Jose D. Cordova II, grandson of Jose Dolores Cordova I, and a third generation miller, saw a copy of the Fall 1982 issue of the Old Mill News at the La Cienega Ranch in New Mexico. After seeing four seventeenth century mills in the model Spanish village, they decided to become SPOOM members.

Impressed with photos and articles about other mills in various stages of restoration they began to think about his grandfather and father's mill in Jarales. After working in the mill during high school, he pursued his milling interest and graduated as a milling engineer from Kansas State University in 1964.

Jose Cordova II estimates over 10,000 hours have been spent transforming the tin and wood frame mill building and the 1920s-vintage equipment into a modern food manufacturing plant. He said it was a labor of love and "the project of a lifetime" to restore the antique equipment for the milling of tortilla flour. Today the mill is fully restored and operational. For more information call (505) 864-0305.

Story and photo courtesy of Kathryn A. Cordova.

The Covell Mills
New York

In 1813, William Covell and wife, Betsy, bought the mill property in South Hartford, New York from Jesse Hills. An operating mill existed here and evidence indicates an up-and-down sawmill. Covell learned gristmill operations from his father, owner of a gristmill in North Hartford. It appears Covell continued operating Hills' Mill but built a gristmill in 1816.

The building of the Covell Mill, a dramatic event, began in the famous "Year Without a Summer", when the region had been deprived of its usual share of sunlight by a volcanic eruption in the South Pacific. Spring did not bring warm weather and local farmers plowed fields wearing winter clothing. A poor crop yield enabled Covell to hire men for digging the necessary headrace and wheelpit from the slate cliff beside the millstream. Some workers walked from Vermont, at least twelve miles away, to dig and earn wages. Each received a peck of corn a day for his labor.

A sturdy one and a half story frame in the classic post and beam style was erected over a thirty foot stone foundation. The interior has double beam braces and handsome wide pine sheathing throughout.

When Betsy died in 1828, Covell married a young widow, Lovicy Whedon. Lovicy learned to operate the sawmill and gristmill, while raising the children. In the 1840s the Covells built a smaller waterpowered mill nearby to produce plaster. This mill, installed on higher ground, obtained water from a new dam much farther upstream. The water moved to the mill in a 500-foot canal that clings dramatically to a hillside. William Covell died in 1843, and Lovicy deeded the plaster mill to son James in 1852.

The interior of the gristmill shows several changes in water-wheel positions were made, obviously seeking greater efficiency. Lovicy installed a steam engine to operate the gristmill and a circular sawmill that replaced Jesse Hills' sawmill after 1855. In 1870 son, William Jr., converted it to a planing mill to produce wood turnings, tool handles, and various other wooden articles.

Lovicy continued to operate the sawmill and gristmill alone after William's death until her own in 1872.

Fred and Joan Patton completely restored the gristmill throughout 1973 and 1976. It is now their home. For more information call (518) 632-5358.

Story and photo courtesy of Frederic and Joan Patton.

JOHN A. DEBAUN MILL

NEW YORK

The following article about John A. DeBaun was written in 1903 in the *New York Sun:*

Feats of a One-Handed Man

A three minutes' walk from the station at Tallman, Rockland County, New York, brings one to a two-story building close by a little street and an artificial pond. Here John A. DeBaun has worked many years, accomplishing what it might seem impossible for anyone to do without a pair of good hands.

Mr. DeBaun was born January 8, 1827. When two years old he suffered a stroke of paralysis, which left his right arm useless. There is a little activity in the hand, which can feebly grasp an object placed in it. Early he manifested a mechanical turn of mind, and when a boy surprised his parents and playmates by making toys. A stock and dyes made out of horseshoe-nail rod with a file, bow guns, matchbook guns, a turning lathe, windmills and waterwheels operating trip-hammers were some of the boy's handiwork.

When a young man he made the patterns for much of the ironwork of a lathe which he has used for fifty years and more. This lathe is made left-handed, especially adapted for his use. He built up a general repairing business, and repaired guns, clocks, pumps, tinware, and the like. He has made many a gunstock, and he constructed a wooden-barreled gun, which has been successfully fired.

Having both a fondness and a good ear for music, it was natural for him to become interested in musical instruments. He first took up the accordion, and for years had a good business in repairing such instruments when they were used more than they are today. About the beginning of the Civil War he started to make melodeons. He did the work himself, even sawing the scroll for the bookracks, putting the ivory on the keys, veneering the case and turring and voicing the reeds. Few men have more completely built musical instruments than has Mr. DeBaun. His melodeons were noted for being well built and full and rich in tone. The organ has displayed the melodeon, and Mr. DeBaun has practically ceased the manufacture of musical instruments.

The printer's realm was next invaded, and for several years Mr. DeBaun did quite a business in the way of printing circulars and labels. Not satisfied with his press, he by ingenious devise converted it into a self-inking rotary press.

About twenty years ago Mr. DeBaun decided to put a set of millstones into his shop and grind his own chicken feed. With no experience in such work he cut down, refaced and furrowed a set of old millstones, remodeled the machinery and built other necessary apparatus. From the first it was a success. Not only was corn cracked for the chickens and the horses, but buckwheat was ground for the house. Soon the neighbors began to patronize the new miller. The business has grown, and new machinery and apparatus has been added from time to time until now general milling in done.

He purchased a small, 20-inch set of unfinished millstones and made them into a rigid underrunner mill. This is believed by him to be the first set of millstones so set up. Since then the method has become quite general. In fact, many devices in all parts of the shop prove Mr. DeBaun to be an inventor. Although he has never taken out a patent, he has seen some of his ideas and devises patented afterward by others.

Besides being engaged in these mechanical pursuits Mr. DeBaun has found time to do some literary work. He has several unpublished manuscripts on religious and metaphysical subjects.

Mr. DeBaun is now 76, straight as an arrow and walks as spryly as a young man of twenty. He is ready to converse on almost any subject from the number of barrels of water a minute that runs over the waterwheel in his mill to the action of the nervo-electric forces in hypnotic phenomena.

New York Sun 1903.

Story and photos courtesy of Joe & Valerie Harvey Rinaldi.

Eden Valley Mill

New York

Eden Valley Mill was built in 1812 on the south branch of the Eighteen Mile Creek in Eden, New York. Pioneer settlers named it "Eden" because of the resemblance to their idea of paradise. In the 1850s the valley was a thriving mill community. More than fifty carloads of feed and flour were ground in a season. Half a million board feet of lumber and thousands of lath and shingles were cut. Innumerable bushel boxes and baskets and approximately 10,000 berry crates were made each year. The gristmill made some of the best buckwheat in New York State. At every falls along the creek, from Lake Erie to the foothills, were thriving mills. The valley was alive with waterpowered industry. Today Eden Valley Mill and two other nearby mill sites make up a New York State designated Historic District

Water ran through an above ground sluiceway and undershot waterwheel to power the wheel. In the 1890s an underground sluiceway and turbine wheel supplied the power. After the dam went out in 1929, power came from a modified Buick automobile engine using natural gas for fuel.

In 1910 a creamery was added to the site. It received milk from Eden Valley farmers, made cheeses and butter and sold excess cheese to a Buffalo food purveyor. A later innovation was the introduction of "Valley Velvet Ice Cream" in cones and various size bulk containers. The ice cream sold at the creamery, the Erie County Fair and was distributed to many retail stores in the area.

In addition to the gristmill and creamery, the cider mill was an outlet for local apples. They were pressed through large fabric covered screens to sieve out the pulp, seeds and peels from the sparkling sweet juice. The cider sold in glass gallon jugs and was eagerly awaited each fall.

In 1990 the Town of Eden obtained the mill and received a state grant and private donations to stabilize the building. The stabilization was completed in 1991 and currently the mill stands as pictured. For more information call (716) 648-1497.

Story and photo courtesy of Gary Eye.

The 1835 Mill

New York

The stone gristmill complex on Timmerman Creek located in the town of St. Johnsville is now known as "INn by The Mill". Built Ca. 1830 and first operated in early 1835, the building exhibits a high level of craftsmanship. Originally powered by a thirty foot overshot waterwheel housed within the building, the gristmill operated as a custom mill under a series of owners until the 1930s. The mill produced stone-ground flour as its principal product throughout the nineteenth century. Converted to water turbine power and roller mill technology in 1898, the mill continued to grind buckwheat and animal feed for local dairy farmers until it ceased operation in the 1930s.

The opening of the Erie Canal in 1825 proved a great stimulus to the local economy of Montgomery County. During the early decades of the nineteenth century, local settlement patterns were largely determined by the numerous streams flowing into the Mohawk River, which became the site of small, water-powered grist and saw mills.

The stone gristmill pre-dates the establishment of the town of St. Johnsville (1838). Built of locally quarried limestone block, the mill was completed by early 1835. An inscription written on a wood panel notes that the building was erected by millwrights Leonard and Curran. The substantial walls and heavy framing were designed to resist torque and vibration from the massive thirty foot diameter, eight foot wide overshot waterwheel. The wheel was installed inside the stone structure to minimize weathering and extend the grinding season. A dam across Timmerman Creek provided waterpower for operation.

The mill was first owned by the Beekman family, who employed Samuel Sadler as miller to operate the flour mill for its first decade. In 1845, Loami Beekman became the miller.

In the cash-poor agrarian economy of the Mohawk Valley, the miller retained a portion (usually one-tenth) of the flour ground as payment. The flour was hauled by wagon to St. Johnsville and shipped to urban markets via the Erie Canal and later the New York Central Railroad. In 1884, Absalom Everett Seaman, acquired the mill for $6,000 and operated it until 1921. Seaman kept a diary, recording dates and details of construction, alterations and business transactions. He erected the existing mill owner's house in 1894, the barn, wagon shed and hog house. By the end of the nineteenth century, dairy farming supplanted wheat growing as the principal agricultural activity in the Mohawk Valley. To meet this change in market conditions, A.E. Seaman replaced the mill's waterwheel with a more efficient water turbine in 1898. Roller milling machinery was substituted for the millstones, to grind buckwheat flour. After milling ceased to be profitable, the stone structure stood vacant and open to the elements for nearly half a century.

The mill is a gable-roofed structure constructed of rough cut, native limestone block laid random ashlar. The building rises from four stories at creekside to three stories at grade. The ruins of the former mill dam remain approximately 250 feet upstream from the mill. The dam created a millpond, from which water was diverted and conveyed in an above ground timber flume to the overshot wheel inside the mill. In 1898, the timber flume was replaced with a twenty inch cast iron penstock to power the newly installed water turbine. Only the stone abutments on both stream banks remain. The stone gristmill, mill owner's house (1894) and ancillary buildings reflect the late nineteenth and early twentieth century character of a rural waterpowered mill site.

Rehabilitated after 1988, the stone gristmill complex is currently a bed and breakfast. For more information call (518) 568-2388.

Story and photo courtesy Ron Hezel.

FRONT STREET MILL
NEW YORK

The first gristmill, built in 1842, was rebuilt using stone in 1846. According to local folklore, the stone mill burned down in 1913 and was rebuilt once more by the two sons of Benjamin Morrison that same year. The present mill construction used beech and hemlock trees from a lot apparently owned by Mr. Morrison. It is believed he hired a wrecking firm to bring equipment by rail to level the original foundation. This new mill had tin siding. To this day you can still read the starting date of the mill, June 1914, handwritten by Frank Morrison on an original beam. Coal sheds were built when Morrison expanded the operation. Herb Wainwright, the last manager of the working mill, began in 1946 and the mill ran until the late 1950s producing bulk feed, grain and coal.

Various owners gutted and sold the mill machinery after operations ceased.

Kyle D. Hartman purchased the mill in 1977 from Albert Hundley, an implement dealer, who had received the mill in trade for a tractor. It is now the residence of Mrs. Priscilla Hartman and has hosted the Rensselaer Falls Historical Society meetings since 1986. The mill site is the location of the Oswegatchie River House Museum and Art School. It is also the home of Ladies First, Gallery of Art. For more information call (315) 344-8882.

Story and photo courtesy Sally and Kyle Hartman.

HANFORD MILLS
NEW YORK

Hanford Mills Museum in Delaware County is the only operational waterpowered sawmill, gristmill and woodworking factory remaining in the Catskill Mountain region.

This rural industrial complex, with its picturesque pond, focuses on the period between 1890 and 1930 when the business was at its height.

Jonathan B. Parris built the original sawmill in 1846. This sawmill went through five owners in quick succession until D.J. Hanford purchased it in 1860. D.J. Hanford added a gristmill in 1869 and his first specialty woodworking machine in 1876. As the mill became a year-round business, the Hanfords continued to add machinery, creating a rural industrial complex. They produced lumber, feed, barrel heads, broom and tool handles, and milk crates. When D.J. Hanford died in 1899, his sons, Horace and Will took over the mill business; a third son, John, took over his father's farm. Horace Hanford sold the mill in 1945 to three employees, Mike, Frank, and Joe Pizza. They continued operating the business until 1967. In that year, Ken Kelso purchased the buildings and opened the site as a private museum. The museum was incorporated as a nonprofit organization in 1973. Listed on the State and National Register of Historic Places, the site includes sixteen historical structures.

In addition to the mill, the site includes several other buildings. The Hanfords built the feed mill to store items brought in on the Ulster and Delaware Railroad and later converted it for feed storage. Today the building houses a theater, and offers exhibits. A boxcar, loaded as it might have been in 1901, sits behind the Feed Mill. Other buildings on the museum site include a lumber shed, horse barn, hardware store and a white barn with an agricultural exhibit.

Adjoining the Hanford Mills Museum site is John Hanford's farmhouse. This house, built in 1909, replaced D.J. Hanford's older home, which stood in the same location. The museum restored the building to the 1920s, when Hanford's farm was in its heyday. The house contains many original Hanford furnishings and an exhibit on John Hanford's occupations. For more information, call Hanford Mills Museum at 1-800-295-4992.

Story and photo courtesy of Caroline de Marrais, Curator.

Dard Hunter Mill
New York

The Dard Hunter Mill was the third mill built on the Gomez Mill House site in Marlborough-on-Hudson. A Sephardic Jew, Luis Moses Gomez, built the 1714 manor house for use as a trading post. A sawmill for the trees cut from the surrounding virgin forest sat adjacent to the main house. In 1772, a gristmill was added along the stream known previously as "Jews Creek" and now "Acker's Creek" or "Mill Creek".

By 1913 world famous paper craftsman Dard Hunter had returned from his studies in Europe. He bought this two-hundred-years-old house and built his mill in the shape of a sixteenth century Devonshire cottage, with a rye thatched roof. He found no evidence of other mills except two millstones, which he placed in the walkway in front of the old house. He nurtured the idea of producing an edition of books wholly unassisted by any person, or finished product. He resolved to purchase neither paper nor type. Everything must be the product of his own mind and skill. He felt the previous books produced by him at the Roycrofters lacked the charm of old volumes. He wanted to produce books with the feeling and texture of those of the Italian printers of the fifteenth and sixteenth centuries, whose work has never been excelled. He brought the equipment for this mill from a small village in Wikshire, England, where he discovered it in an ancient paper-mill.

Hunter reduced linen rags to a fibrous pulp with the aid of an old, creaky, wooden waterwheel and made every sheet of paper separately in a hand-mold, using exactly the same methods employed by craftsmen four hundred years ago. With no tools or utensils other than those used by type cutters of the days of Aldus and Jenson, he designed and cut a font of type, struck the matrices, and cast the type in a hand-mold. When he had sufficient waterpower he turned out approximately 150 sheets of paper a day. When the waterpower dried up he cut and cast the type, about one hundred thousand pieces, needed for his project.

Peter Thomas summed up the historical importance of the mill as follows, " ....for the first time an individual, rather than an apprentice-trained crew, made sheets of paper by hand — all aspects of the operation being completed by the same individual. This break from the traditional method of producing handmade paper paved the way for the modern renaissance in hand papermaking. If hand paper-makers had holy places to pilgrimage to, this would surely rank highly among them."

The Dard Hunter Mill has been purchased by The Gomez Foundation, current owners and descendants of previous owners. Restoration of the mill wheel is currently underway. For more information call (914) 236-3126.

Story and photo courtesy of William Maurer, Director.

The Log Village Gristmill

New York

The little town of Hartford, Washington County, New York was still in its infancy when Hezekiah Mann built a gristmill on the banks of East Creek. The hamlet became known as Log Village because most dwellings were made of logs. Local records indicate that the mill construction took place around 1810. Little is known about Hezekiah but he was no ordinary carpenter. His knowledge of mechanics and hydraulics put him in a class with his contemporary, Oliver Evans, a well-known millwright of that time.

The first white settlers came to Log Village, now known as East Hartford, by 1781 or 1782. At the time of the building of the mill, much of the land had been cleared and log homes erected. Other needs such as sawmills, tanneries, stores, churches, blacksmith shops and schools were being provided for the pioneers. Hezekiah's creation served the community continuously until 1902.

Nationally, James Madison, our fourth president, was having considerable difficulty with the English and French who interfered with American shipping. Such activities culminated in "Mr. Madison's War", the War of 1812. Had there been newspapers then they would have been full of stories about Thomas Jefferson, Henry Clay, Andrew Jackson, Daniel Webster, John Calhoun or John Quincy Adams. Kentucky, Ohio, Tennessee, and Vermont had just been granted statehood adding to the original thirteen to make our United States seventeen in number. Lewis and Clark returned from their expedition. Abraham Lincoln celebrated his first birthday. Fulton completed his successful steamboat run from New York to Albany. Yet to come were the Erie Canal, Stephenson's locomotive, Morse's telegraph and the discovery of oil.

Floyd and Mildred Harwood purchased the mill in 1972 and after five years of restoration started grinding again. The mill had been idle for seventy-five years and required the building of a new dam, a seventeen-foot wooden waterwheel, six feet wide, a sixty-five-foot wooden flume, thirty inches in diameter, and three foundation walls. It has three stones, two french buhr and one granite. The eighteen-foot bolter is used to sift the wheat and buckwheat flour. The second floor contains a nineteen-foot wood lathe, and other equipment for making wooden patterns for castings.

The Harwood's continue to operate the Log Village Gristmill grinding corn meal and whole wheat flour. For more information call (518) 632-5937.

Story and photo courtesy Floyd Harwood.

New Hope Mills
New York

New Hope Mills Inc. has produced flour for over 175 years to earn distinction as one of America's oldest continually operating flourmills.

Judge Charles Kellogg built the flourmill in 1823. He was a member of the New York State Assembly (1808-1809; 1821-1822) and later elected to Congress (1825). Judge Kellogg sold to Horace Rounds in 1851. The mill remained in the Rounds family until 1947, then sold to Howard Weed, Hubert Latta, and Leland Weed and became New Hope Mills. Leland Weed became sole owner in 1953. Today, the mill is owned by two of his sons, Dale and David Weed.

This mill and a sawmill, a short distance upstream, are known as New Hope Mills Inc. There were fifteen waterpowered industries along the banks of Bear Swamp Creek in the 1850s. Only these two mills remain in operation.

Local history contends that in 1823, after the placement of the mill's forty-foot, five-sided ridge pole, a man climbed to the top and threw a gallon jug of whiskey over his shoulder declaring the town of Sodom to be renamed New Hope.

The mill originally contained three runs of stone and had a capacity of 200 bushels a day. In 1892 the mill began grinding with the roller mills. In 1986 installation of a pneumatic mill took place, the first of its kind.

New Hope Mills has been a waterpowered industry from the beginning to the present. The most visible means of waterpower is the overshot waterwheel brought from New Jersey in 1972. The wheel, about 100 years old and very rusted, required new buckets. In 1978 the buckets were rebuilt of sheet metal that would have been used in making Brockway fenders. It is jokingly known as the only Brockway waterwheel in the world.

The real power comes from the water turbine underneath the mill. The turbine can run all year round providing there is sufficient water. There is an upper millpond with a twenty four-foot waterfall before the sawmill and before the flourmill is a lower millpond with a twenty-eight foot waterfall.

Even today as industries strive for more modern machines and methods to increase production and meet the demands of our society, New Hope Mills continues to operate much the same as it did in the 1800s. The owners of New Hope Mills take pride in using such an old but efficient piece of our American heritage. For more information call (315) 497-0783.

Story and photo courtesy of Dale Weed.

Philipsburg Mill

New York

The Philipsburg Mill is a modern building on the Philipsburg Manor property in Sleepy Hollow. It has been built in the same fashion and on the same location as the original mill constructed by Fredrick Philipse in 1682. It houses two sets of millstones imported from France. The dam is also reconstructed in the same style and location as the original dam. The mill is powered by water from the Pocantico River.

The forests and swamps that blanketed the northeast were highly conducive to milling. Like a sponge, they retained water releasing it gradually into streams and creeks. The steady supply of water kept the mill wheel turning most of the year. Among European settlers in North America, manual labor in short supply created a great demand for labor-saving mills. The mill at Philipsburg Manor was run by enslaved Africans. More than twenty-three slaves are known to have lived and worked there. Their compelling story is told at the historic site.

Since clearing the land was the first task confronting the early colonists, the first mills were primarily sawmills. As farms replaced forest, two other types of mills became widespread. Gristmills ground the grains (wheat, rye, oats, and barley). The same stones were used to grind dry spices, cocoa, pigments and gunpowder. Fulling mills used the waterwheel to drive large mallets to pound wool cloth. Soaking and beating the cloth in this fashion tightened or "pre-shrunk" the woolen fibers, a necessary step before it was sewn into garments.

Construction of most seventeenth and eighteenth century mills involved damming a stream or brook to create a millpond. This guaranteed a sufficient supply of water, funnelled to the water wheel via a spillway or flume. The most powerful type of wheel, requiring the largest dam was the overshot wheel. The flume carried the water over the wheel dropping it into the wheel buckets on the other side. At mills with a breast-shot wheel, the water was deposited into the side of the wheel. When the flume funneled the water to the bottom of the wheel, it was known as an undershot wheel, the least efficient type.

Two other power sources were less used. Tidal mills were built along coast lines. At high tide water backed up into an estuary or pond. A dam built across these bodies of water had a gate that allowed water to rush in at high tide. When the tide began to recede the reversed flow of water caused the gate to close. Twice a day (after each high tide) the miller had enough water to drive a waterwheel for about five and one half hours.

The Philipsburg Mill has French buhrstone millstones fifty-four inches in diameter with a thickness at the center of twelve inches and at the skirt of nine and one half inches. The running stone weighed 2,400 pounds and the bed stone 1,700 pounds. The waterwheel is eleven feet in diameter and fifty-nine and one half inches in width. The flume that carries the water to the wheel is fifty-five feet long, four and one half feet wide and four feet deep. The original mill ground 30,000 pounds of wheat flour per week.

The Philipsburg Mill is now part of the Philipsburg Manor property, which includes a museum, restaurant, gift shop, book shop and boat trips. For more information call (914) 631-8200.

Story and photos courtesy Historic Hudson Valley.
Photo by Mick Hale.

Rest Place Gristmill

New York

The Rest Place Gristmill is hidden away in a secluded spot west of the Hudson River in the Rondoot Valley, between the Schawangunk Ridge and the Catskill Mountains.

This is one of a few rural historic districts and, according to Sylvester (1875), among the oldest in Ulster County, New York. Over the door is written "Built 1765 1846." This is cut into a single piece of stone. The second date is around the time of the Industrial Revolution. The present heavy timber frame building contains two pairs of French buhrstones, despite local quarrying of Schawangunk conglomerate. On the plaster cover of one of the buhrstones is indented "Vote for Alton B. Parker" an Esopees contender for President against Teddy Roosevelt in 1904.

Rest Place, or Reust Plause on the early deeds, served as a stop over at the ford on the stage route halfway between Kingston, the county seat and first capitol of New York State, and Ellenville on the Old Mine Road, originally an Indian trail, to the Delaware River Valley. Nearby, one part is still known as The Old Kings Highway.

It is believed the builder, Wessel Brodhead, grandson of Capt. Daniel Brodhead, was with the English fleet that occupied New Amsterdam in 1771. Wessel willed the mill to his two sons, Charles and Lewis. The first known miller was Abraham Sahler, Jr. from Perkiomen, Pennsylvania. He married into a local family in this closely knit historic area of German Palatinate, Dutch, French-Huguenots, and English. Their fields were guarded day and night against the Indians and Tories during revolutionary times.

In 1858, the mill sold, and for many years was known as the Cyrus DePuy Gristmill. Cyrus built a bridge and acquired land for a lane to the Kyserike Road, now Rest Plause Road.

The Delaware and Hudson Canal (1828-1885) brought in bulk farm needs from Pennsylvania. The Ohio and Western Railroad, (locally known as the Old and Weary) ran through the property and replaced the slower barges. Around 1914 the mill business slowed, then ended when a flood took out the original dam.

Clarence Hansen began restoration in 1961. Charlie Howell, newly arrived Master Miller at Philipsburg Manor, provided outstanding assistance when Cliff Sheeley began renovations with a new dam at the falls, sluiceway and ten foot overshot waterwheel. Iron gears in the original husk supported the stones.

When searching for cypress at small Florida sawmills, Clarence's brother found Arthur P. Pfeil of Buffalo. He had century old boards from out-dated beer vats. The wheel was cut there and assembled at the mill site by Cliff Sheeley. Cleaners and sifters were found on the top floor and cocks loft; missing chutes and belts were replaced and once again the gristmill was operational.

There is a Dutch barn on the premises and the stone miller's house is a private residence.

Story and photo by Mrs. Clarence Hansen.

Shinglekill Falls Gristmill

New York

Shinglekill Falls Gristmill, once known as Wright's Mill, is the last remaining of more than fifty waterpowered mills developed in Greene County. The building, constructed in 1894, replaced a nearly identical structure that burned in 1892. An iron forge, a factory that made wooden shingles, and an earlier gristmill burned around 1850.

At thirty-two feet, Shinglekill Falls is the largest of a series of waterfalls. Each powered at least one mill along that short length of stream.

Other mills on the Shinglekill included a cooperage that made barrels and buckets, a sawmill, furniture factory, a factory that made wooden clocks, tanneries, a succession of other gristmills and a wood turning mill that initially made farm implements.

In the 1800s, Alpheus Wright purchased, from an uncle, an up and down sawmill, located at the next waterfall upstream. He then built a furniture factory on the other side of the stream. Later he set up his son, Sylvester, in the gristmill on the big waterfalls and his other son, Anson, in the wood turning mill just downstream. Changing agriculture and the miller's fondness for alcohol resulted in very little use of the gristmill. It ceased operation in the early teens when freezing of the spring seepage from under the building caused the dry laid stone foundation under the massive twenty-eight foot diameter by six foot wide wood and cast iron overshot waterwheel to collapse.

When Anson Wright took over the turning mill in the late 1800s, there was no longer a demand for the grain cradles, tool handles and other farm implements that had been its mainstay at the time of the Civil War. He used the building to make souvenirs to be sold wholesale or by consignment. The gristmill building eventually became a retail store for the souvenirs. Only the foundation and the cast iron parts of the waterwheel remain from the turning mill.

In collaboration with a photographer, who had a studio beside the falls, picture post cards of the Catskill Mountains were added to the line of souvenirs.

Admission to the Shinglekill Falls was ten cents. This allowed a swim in the chilly water, sunbathing on the rocks and a drink from the foul smelling sulfur spring that was touted to be a fountain of health and potency. The atmosphere could provide romatic inspiration to steal a kiss or more and photographs could be taken in front of the falls or the mighty waterwheel. Trout fishing was among the activities, and souvenirs, candy and soda could be purchased from the store.

Some things have not changed much over the years and others are gone forever. The old wheel was sold for scrap metal in 1939. The current owners are descended from the Wrights. Restoration and installation of a twenty-two foot Fitz wheel and converting the miller's house to a bed 'n' breakfast are among their planned retirement projects.

The morning sun produces a rainbow in the spray around Shinglekill Falls but the late day sun sheds the best light on the mill. For more information call (518) 864-5068.

Story and photos courtesy of Tracy Lamanec.

Stony Brook Gristmill
New York

In colonial America, the sturdy gristmill became the hub of community life. There farmers gathered to have their grain ground into flour, exchange local news and transact business. As the largest buildings in rural areas, the mills with their roaring machinery were an exciting place to the young and of never-ending interest to the old.

Adam Smith, son of the legendary Richard (Bull) Smith, built the Stony Brook Gristmill in 1699. The original deed read: "Adam Smith shall have the town's right of the stream called Stony Brook, with two acres of land adjoining thereto which may be most convenient, on condition that he erect and build a good sufficient gristmill and maintain the same." After the original mill washed out, the present structure replaced it in 1751.

William Davis, who owned the mill in 1846, not only ground grain but also sawed logs using waterpower. Prior to 1860, owners of a mill, shipyards and smithy were considered the most important proprietors.

In 1885, a German, Edward Kane, purchased the mill, smithy and wheelwright shop, as well as the hilly land to the west. He planted vineyards on the island in the middle of the pond and the grapes were pressed and fermented in this old mill. A jug of wine had been sealed in the foundation when the mill was enlarged.

Kane, no miller, employed the services of a Bohemian, Alois Kopriva, a graduate of Prague University in Czechoslovakia. He and his son, Alois Jr., were schooled in the art of milling. He, undoubtedly the best miller the town ever had, introduced the process used in the mill today. In addition, Alois played the violin, entertaining his guests with gypsy music. At the turn of the century Alois Jr. operated the mill for owner, Temple Emmett. He taught Frank Schaefer, the last miller. Frank said Alois Jr. would play his violin for hours to the accompaniment of the turning waterwheel. Miller Aobert Grassi dressed the millstone. The daughter of Alois Jr. commented recently that her father sold wheat germ to local physicians because they thought "wheat germ was better than surgery."

As recently as the 1950s, farmers still brought wheat and corn to be ground. Miller, Frank Schaefer ground most of the grain or sale as natural flour. Schaefer said he seldom used the bolting machine because most of his customers wanted the natural article with the beneficial wheat germ. Natural wheat is a perishable product, so it had to be ground daily and was shipped in small quantities to users in forty-two states.

After acquiring the mill from the Emmetts in 1947, philanthropist, Ward Melville retained architect Richard Haviland Smythe to undertake exterior renovations on the structure. He then deeded the gristmill to The Ward Melville Heritage Organization for preservation.

Although repairs had been made over the years, the gristmill remained structurally unsafe and closed for restoration in 1988. In 1990 the gristmill gained designation on The National Register of Historic Places and New York State Register of Historic Places as a rare example of eighteenth century Dutch farming.

In the late 1960s under lease agreement with The Ward Melville Heritage Organization, The Museums at Stony Brook reopened the mill. In 1987 The Ward Melville Heritage Organization again took over the operation of the gristmill. For more information call (516) 751-2244.

Story and photo courtesy of The Ward Melville Heritage Organization. Photo by Richard Rocchio.

Bull Pen Mill
(Jackson County), NC
Photo by J.C. Barger

Located on Fowler's Creek (a major tributary of the Chatooga) in Cashiers, North Carolina, Bull Pen Mill rests beside a forty foot waterfall.

The mill's origins are obscure but early ancestors of the McCalls, a leading family since the area was first settled, probably built it in the 1830s. Wack McCall is shown as the earliest established operator, during the years 1870-1890.

Bull Pen Mill began as a tub mill. A tub mill differs from the more familiar and efficient overshot waterwheels in that it is seated horizontally, and the flow of water pours onto its surface and drives it in a fashion like today's turbines.

A wooden overshot waterwheel first replaced the original tub mill. Dr. Fancher, who restored the mill in the 1940s, replaced this with the metal overshot wheel still in use, and a metal flume. The current owner, Forrest Cole, who has most recently restored the mill, hopes to eventually have the metal piping covered with a wooden sheathing, for greater authenticity of appearance.

The original mill may have served all of the families in the nearby Pleasant Grove Community, which numbered as many as fifty persons by the turn of the century but then dwindled almost to extinction when most of the families moved out during the 1935-1940 period. Only one family of McCalls remain from among the earlier settlers. In recent years a number of new houses have been built not too far away, and the process continues.

From the earlier mills on the site, Cole has preserved the original grindstones, central iron shaft, and original hopper, as well as the 1940s waterwheel and gearing. By Lee Copple, The Highlander, Highlands, N.C. June 29,1978.

Courtesy Society for the Preservation of Old Mills.

Bull Pen Mill
(Jackson County), NC
Photo by J.C. Barger

Brookshire Mill
North Carolina

Hawards Creek (Near Boone), NC
Photo by Joe Seme

Floyd's Mill
North Carolina

Swain County, NC
Photo by J.C. Barger

High Hampton Mill
North Carolina

Cashiers (Jackson County), NC
Photo by J.C. Barger

Hoggard Mill
North Carolina

Windsor (Bertie County), NC
Photo by J.C. Barger

Lick Log Mill
North Carolina

Highlands, North Carolina
This one-man gristmill was built in 1874 from hand-split chestnut logs and has wooden cog machinery.
Photo by Alden Miles
Courtesy Society for the Preservation of Old Mills.

Linney's Mill
North Carolina

Rocky Creek winds through the northeast corner of Alexander County into Iredell County and powered a total of eleven mills. Linney's Mill is one of the few remaining and the only one continuing to operate by water-power.

Richard Cook first established the mill in 1790. This mill washed out in the 1917 flood. W.T. Linney bought the mill site in 1936. At the time of restoration only a small part of the original dam remained and there was no milling equipment.

William and Grace Linney operated the mill from 1945 to 1992. In 1992 his son and daughter-in-law, Billy and Juanita Linney took over the operation of the mill. The mill is driven by an impressive nine foot by thirteen foot overshot waterwheel, which reportedly produces forty-five horsepower. This wheel turns the vertical thirty inch meadows millstones and grain cleaning machinery. It also turns an electric generator at other times to produce the power to drive feed milling machinery. There is a swinging bridge below the mill 110 feet long over the creek and a campground. For more information call (704) 592-2075.

Story and photo courtesy of Billy Linney.

Little Mill
North Carolina

Little Mill
From Henson Creek
Linville Falls (Burke County), N.C.
Courtesy Society for the Preservation of Old Mills

Little Mill
Jonas Ridge (Burke County), NC
Photo by J.C. Barger

Mill House
North Carolina

Mill House, located at the John C. Campbell Folk School in Brasstown, North Carolina, is one of the earliest and perhaps most outstanding buildings on the campus. Its architecture has strong European influence and reflects the different formative influences and functions of the school. Leon Deschamps, a Belgian who was the school's engineer and forester, designed and built Mill House in 1928. It housed the first wood shop on the lower level, with power supplied by a drive shaft propelled by the mill wheel.

The wheel also powered a pump, which sent spring water up a mountain to a reservoir from which a gravity water system could flow to the whole campus. Later the wheel turned a generator, which powered the first electric lights in the area. This string of bulbs lit the dairy for early morning milking.

Male students were initially housed in the dormitory on the second level. When the new woodworking shop was built, the lower floor was used for a laundry with a large boiler. Still later, it was renovated into a two-story apartment in which many staff members have lived throughout the years. Mill House is now used to house students who are attending classes at the Folk School.

The John C. Campbell Folk School sits on 372 acres in the Smoky Mountains of western North Carolina. It has offered classes in arts, crafts, music, dance, nature studies and cooking since 1925 and is listed on the National Register of Historic Places. For more information, please call 800-365-5724.

Story courtesy John C. Campbell Folk School.

Mattawan (McDowell County), NC
Photo by J.C. Barger

Mingus Mill
North Carolina

Cherokee (Swain County), North Carolina
Photo by J. C. Barger

Mingus Mill
Cherokee (Swain County), NC
Photo by J.C. Barger

Mingus Mill
North Carolina

Cherokee (Swain County), NC
Photo by J.C. Barger

Moravian Falls
North Carolina

Moravian Falls (Wilkes County), NC
Photo by J.C. Barger

Morgan Mill

North Carolina

Morgan Mill, built in 1855, is located in Brevard, North Carolina.

Transylvania County history indicates that the mill was built by one of the county's early pioneers, named Morgan, who went west to dig for gold to stake the mill.

After the War Between the States, he returned to the Cherryfield section of the county and built the mill. He used twelve inch, hand hewn chestnut logs secured with hickory and locust pegs, brought in millstones from France via sailing ship and wagon train, and began grinding corn, rye and wheat. The old mill is still grinding today. The original wooden wheel has been replaced with a thirty foot iron mill wheel.

Courtesy Society for the Preservation of Old Mills.

Murray's Mill
North Carolina

The Murray's Mill Historic District in Catawba, North Carolina provides a glimpse back to an era integral to the industrial and cultural heritage of rural Catawba County. Surrounded by the rolling hills, woodlands, and picturesque pastures typical of the North Carolina Piedmont region, the complex is comprised of nearly two dozen milling related structures.

It is the centerpiece of the district is the mill, enlarged in 1913 by John L. Murray; built on the site of the original mill erected by his father (William) in 1883. Murray's Mill was a vital facet in the rural agrarian economy of Catawba County, operated by four generations of the Murray family. The mill remained productive and profitable through two World Wars and the great Depression. The machinery used in producing flour, meal, and other grain products rest where they fell silent in 1967, when Lloyd Murray ceased mill operations. Today a twenty-two foot overshot waterwheel churns water passing through the flume, which is connected to a concrete dam. Corn milling demonstrations, using the original original French buhrstones, are the highlight of the museum and an annual festival.

Murray & Minges General Store, once the area primary retail outlet, offers insight into part of the mill complex's history and the importance of the store to the local farmers. The Catawba County Historical Association publications and period items are now found on the shelves, replacing the stacks of clothing and dry goods once purchased by farmers and their families.

Resting on the slope adjacent to the mill, the Wheat House has been transformed into an exhibit hall.

North of the millpond is the John L. Murray House, once occupied by John Murray, his wife, and ten children. Period furnishings surround the parlor, and a large stove dominates a basement area used for canning and household tasks.

Murray's Mill is the only surviving mill in Catawba County. For more information please contact the Catawba County Historical Association at (828)465-0383.

Courtesy Society for the Preservation of Old Mills.

THE OLD MILL OF GUILFORD
NORTH CAROLINA

In 1767, before the colonists decided to seek independence from England, Daniel Dillon built a small tub mill in North Carolina on Beaver Creek, and went into business grinding grain for the early settlers. In 1755 the Earl of Granville had granted Dillon the tract of 552 acres on Reedy Fork and Beaver Creeks. Rowan County later granted him a license to build the mill in 1764. Rowan County Court minutes of January 10, 1764 read: "Ordered that Daniel Dillon have License to Build a Public Gristmill on the Reedy Fork of Haw River at the mouth of Beaver Creek." The mill was located in what is now Guilford County, which was created out of Rowan County in 1771.

On February 10, 1781, during the Revolutionary War, British troops under General Cornwallis marched past the mill in pursuit of General Green, encamped at Guilford Courthouse. Legend has it that British troops seized the mill to grind grain for the soldiers prior to the Battle of Guilford Courthouse on March 15, 1781.

The original tub mill and dwelling, on 175 acres, sold to Joel Sanders for $900 in 1808. In 1819, Sanders moved the mill 500 feet down stream and built a new dam across the creek, which increased the millpond to ten acres. The new mill was designed as a merchant corn and wheat mill with an overshot wheel to replace the small tub mill.

In 1913, K.L. Hendrix purchased the mill. He later converted it to a roller mill and replaced the waterwheel with a turbine. In 1932 a state highway was built between the dam and the mill. To keep the mill in operation, the long overhead wooden flume, which carried water from the dam to the mill, had to be replaced by a twenty-six diameter steel pipe to run under the new road. Clarence E. Bailes purchased the mill in 1954, removed the roller mill machinery and replaced the turbine with a twenty-four foot by four foot Fitz overshot waterwheel still operating today.

In 1977, the mill sold to the current owners who continue to operate the mill on a full-time basis. For further information, please call (336) 643-4783.

Story courtesy of Charles and Heidi Parnell.
Photo by J.C. Barger.

Redmond Mill
North Carolina

Statesville (Iredell County), NC
Photo by J.C. Barger

Roger's Mill
North Carolina

Moravian Falls (Watauga County), NC
Photo by J.C. Barger

Sheppard's Mill
North Carolina

Calla Hill Sheppard built Sheppard's Mill in Stokes County, North Carolina on Snow Creek in 1904. The mill replaced an earlier gristmill on nearby Ugly Branch. A two-story weatherboarded frame structure with a gable roof, the mill has a long one-story shed section used as a saw and planing mill. Three turbines provide power for the roller mill machinery, two sets of stones and the sawmill.

The mill produced cornmeal, rye flour, hush puppy mix, plain flour and feed, in addition to the operation of the sawmill. Because of the varied operations at Sheppard's Mill, it was alternately called the Snow Creek Roller Mill and the Snow Creek Woodworking Plant.

Adjacent to the mill are a store and a warming house, where people kept warm in the winter while waiting for their flour. Both structures are still standing, but have since been converted to other uses.

The mill operated until the early 1950s when Sheppard died and Hurricane Hazel washed away the wooden dam. In the 1960s the dam was rebuilt with concrete and the mill totally restored but never put back into operation. The mill is currently being refurbished by the present owners who purchased the mill in 1994.

For further information, please call (336) 874-2777 or (336) 643-4783.

Story courtesy of Charles or Heidi Parnell.
Photo by J.C. Barger.

Wyatt Nance Mill
North Carolina

Asheboro (Randolph County), NC
Photo by J.C. Barger

YATES MILL
NORTH CAROLINA

Yates Mill (circa 1756) is located in Wake County, south of Raleigh. It is the last remaining mill of some forty mills that once operated in the Wake County area and one of a few remaining examples of early American milling in North Carolina.

Yates Mill operated as a waterpowered gristmill for over 200 years grinding corn, flour, sawing logs, and carding wool. The mill structure and the equipment show the evolution of early industry in the lower Piedmont region of North Carolina. In 1963, North Carolina State University purchased the site from A.E. Finley. Yates Mill has been the object of study and preservation efforts since 1970 and was placed on the National Register of Historic Places in 1974. From that time until 1989 the mill stood neglected.

Yates Mill Associates, Inc. began to assist in the restoration, preservation and eventual operation of Yates Mill in 1989. The mill structure has been stabilized, exterior walls, floor and wall beams, studs, rafters, flooring, window jambs, sills and siding have been restored or replaced. Restoration was interrupted in September 1996 by Hurricane Fran. Floodwaters washed out the dam and underpinnings of the mill causing the collapse of the sawmill shed.

Yates Mill was one of the most severely damaged historic structures in North Carolina. If not for prior work accomplished by Yates Mill Associates, it is likely the mill would be gone. Restoration efforts continue despite the setback. The dam and the sawmill shed portion of the mill have been restored. The next step is installation of the rebuilt water wheel and restoration on the interior equipment.

When completed, the working mill will be the focal point of a public park being developed at the site. Yates Mill Associates formed a partnership with Wake County and North Carolina State University to create a historic and environmental education park at the site. In addition to the working eighteenth century mill, the park will contain an interpretive center, field classrooms, and educational trails through a 612 acre nature preserve. Creation of the park will protect significant and diverse ecological habitats. Completion of the project is anticipated in the year 2001.

Story courtesy of John G. Vandenbergh.
Photo courtesy Society for Preservation of Old Mills.

Bear's Mill

Ohio

Milling in Ohio first began around the time of the Greenville Treaty, 1790, with the first mill built in Washington County near Marietta. The first gristmills were constructed under hazardous conditions with the constant threat of Indian attacks. However, settlers needed the mills so urgently that many persisted in building the mills despite the abundant danger. Not until General "Mad Anthony" Wayne defeated the Indians did the construction of mills keep pace with the advance of pioneer settlements, and millers were available to safely grind corn into meal and wheat into flour.

Built in 1849 by Gabriel Baer, Bear's Mill is one of the last operating waterpowered mills in Ohio today. The site of the present mill and water rights, were granted to Major George Adams by President James Monroe in 1824. Bear's Mill was placed on the National Register of Historic Places in 1975.

Another mill flourished on this property prior to 1848 when Baer, a prominent Pennsylvania miller, purchased the land. At the time Moses and Manning Hart had already started construction of the present mill. The framework and siding on the building are original. The hand hewn timber framework is exquisitely done with beams ranging from twelve inches by sixteen inches and fifty feet long without a single splice. The siding is American Black Walnut. The central bay has grain-handling doors on all four levels serviced by a waterpowered wench with cable from the peak. This is how the grain is moved upstairs.

Water from the creek still provides power that turns the underwater turbines to power the milling machinery. A large and small dam create a two-mile lake area from which water is channeled to the millrace. A pair of water gates enable the stream water to be diverted into the race to impel the turbines. A second outlet gate (located under the mill) enables the race to be drained when it is necessary to perform maintenance work on the turbines.

Baer and his son, John, and his son in-law, Emanuel Hershey, completed the mill and millrace. The millrace, 800 feet long, twenty-five feet wide, and ten feet deep, was hand dug by school children who were paid fifty cents a day. The mill became operational January 1, 1850, with Hershey in charge. They continued to run the mill about seven years, perhaps longer. In 1862, Jesse Tillman and John Townsend bought the mill for $8,000 but ran it only a few years, fearing the Confederate soldiers would invade the state during the Civil War and burn the mill.

The millstones are the heart of Bear's Mill. During the heyday of gristmills, a pair of good stones could cost as much as $6,000. They were treated with care and never allowed to become dull or out of balance, as either of these faults would cause premature wearing of the stone. It was not uncommon to require a ten year apprenticeship to learn the art of sharpening millstone. It took a two year journey to France for Gabriel Baer to bring the three imported French buhrstones to Bear's Mill. The French buhr were much sought after by old world millers for their abrasive and porous qualities, resulting in a slow, cool grinding process. This process prevents unnecessary destruction of valuable nutrients, something modern milling cannot claim.

In the late 1860s and early 1870s, the mill passed through many hands. In 1884 the John Stoltz family sold their share to his partner's son, Marcellus Cromer. M.R. Cromer operated the mill longer than anyone has to date, sixty-five years in one family until his death in 1947. Mr. Cromer must have been a colorful character as stories heard from old timers still abound about him!

Waterpowered gristmills in the state numbered almost 2,000 at one time. Today there are only seven working mills left in Ohio.

The current owners of Bear's Mill are Terry and Julie Clark. Terry, a Vietnam Veteran, and Julie, a potter, purchased Bear's Mill in 1979. Terry, having a fascination for old buildings and their workings, was immediately drawn to the mill. Following the purchase of the property, the couple spent three additional months renovating, repairing, and cleaning the mill and grounds to make the mill operational again. In February of 1981, Bear's Mill reopened its doors. For more information call (939)548-5112.

Story and photos courtesy Diana Johns.

Clifton Mill
Ohio

In 1802 on the bluffs of the beautiful Clifton Gorge and Little Miami River Owen Davis built the Clifton Mill. This mill was originally known as Owen's Mill.

The mill witnessed the boom of the village as it became an official stop along the stage coach trail in the 1820s to the 1830s. It provided flour, meal and materials for those traveling west and supplies to those going to the Civil War and the War of 1812.

Most families that owned the mill ran it for several decades. Included were Owen Davis who ran the mill for almost thirty years, John Patterson for twenty years, the Armstrong family for over twenty years and the Preston family for over thirty years. The current owners are the Satariano family who have owned and operated it for almost fifteen years.

Clifton Mill has its original run of stones, forty-eight inch granite and almost two tons in weight. It appears the stones date back to 1802 because of the construction. The stones fit like jigsaw puzzle pieces, granite blocks cut to shape and fastened with heavy bands of iron. Speculation is that in 1802 no quarries existed to make mill grinding stones. These stones were perhaps once the ballasts of supply ships coming into the area.

The mill suffered several fires, the most notable in the late 1880s when part of the mill burned and was then rebuilt on the same site. Originally, the Clifton Mill used a fifteen foot undershot waterwheel that was later replaced by double-sided Leffel turbines.

In the mid 1800s Clifton had as many as five mills. Of these, the Clifton Mill is the only one standing today and operational, reminiscent of Ohio's earliest industry.

For more information please call (937) 767-5501.

Story and photo courtesy of Tony Satariano.
Photo by John Ford.

Indian Mill
Ohio

Indian Mill, built in 1861, is on the Sandusky River. It is the nation's first educational museum of milling in an original mill structure.

The original gristmill and a sawmill were built upstream from this location by the U.S. government in 1820. The name Indian Mill was chosen in appreciation of the local Wyandot Indians' loyalty during the War of 1812. Government-appointed millers ground flour and cornmeal for the Wyandot Reservation. The mills and present structure were owned by a succession of millers after the Indians left.

One miller ran the grist and sawmills to avoid conflict in use of the water supply. The gristmill was primarily used following the harvest.

The two original mills were gifted to the Wyandots, the only action of that type by the government. William Walker, Sr., sub-Indian agent on the Wyandot Reservation used his influence to have the mills built and Charles Garrett (Walker's son-in-law) became the first operator.

George Myers purchased the mills and dismantled the old gristmill. Rummel, a millwright, built the present structure. The mill went through various owners until John Finkle bought the gristmill site of two acres. Finkle's family operated the mill for more than fifty years.

Ward Walton purchased the mill in 1943 and gave it to the State of Ohio under the administration of The Ohio Historical Society. Consultant for the restoration, D.W. Garber of Perrysville, loaned many exhibits to the museum. The museum opened in 1968.

Beams from the original small one-story gristmill, built entirely of walnut, were incorporated into the present three-story structure framed of oak. The mill has covered bridge type construction. Cross trusses secured by steel rods and bolts forged on the site were used to prevent vibration from shaking the structure apart while grinding was in process.

Typically in this type of mill, grain from storage bins on the upper floor feed through a tapered chute into the hopper above the capstone. Waterpower turning the turbine shaft causes the capstone to revolve and breaks and grinds grains into meal and husks. These are then forced into the furrows and carried by centrifugal force to the outer edge and dropped into containers.

Insufficient water now prevents operation of the Indian Mill.

Courtesy Society for Preservation of Old Mills.

Lanterman's Mill
Ohio

Lanterman's Mill in Youngstown, Ohio, rises from the rugged boulders and soft mists of an ancient hemlock lined gorge, majestically towering over a natural twenty-three foot waterfall. The massive fourteen foot overshot oak waterwheel is housed inside the stone and frame structure and powered by the waters of Mill Creek.

Brothers-in-law, German Lanterman and Samuel Kimberly, built the mill 1845 -1846. This is the third mill to occupy the site. The first, a saw and gristmill built in 1798, constituted the earliest industrial-commercial operation in the area. In 1823 a larger mill operated by Eli Baldwin and his son, Jesse, replaced the structure. A flood destroyed the second structure twenty years later. One grinding stone is still visible and rests far down the creek bed as a silent reminder of the violence of that torrent.

Lanterman's Mill was abandoned in 1888. The advent of new technology and development of more commercially viable merchant mills are likely the cause. The building and adjacent land lay neglected until the Youngstown Township Park District (now the Mill Creek Metropolitan Park District) bought them in 1892. The building has been used at various times for storage, boat repair facilities, and a bath house. In 1933, it became the Lanterman's Mill Museum, a trail side nature museum.

Suffering slow but continued deterioration over time, the mill's gradual resurrection began with inclusion in 1976 on the National Register of Historic Places. In 1982 a $600,000 grant from the Florence and Ward Beecher Foundations enabled the park district to stabilize the structure and to install machinery, grinding stones, and a new waterwheel. Finally in 1985, almost one hundred years after the mill's rumbling song was silenced, Lanterman's Mill reopened as a working gristmill.

The waterwheel, huge granite stones, bolting machines, gears and elevators are easily seen during grinding. There is a gift shop on site and nearby a nature trail and boardwalk lead into the cool, rock-walled gorge. There is also a view from the nearby covered bridge, a replica built in 1988 at completion of the site restoration. In 1991 the Park District hosted the national conference of the Society for the Preservation of Old Mills. For more information, call (330) 740-7107.

Story and photo courtesy Tom Mechling, Ginny Elser and Tom Bresko, Mill Creek MetroParks.

Isaac Ludwig Mill

Ohio

Providence Metropark, in Grand Rapids, is home to the historic Isaac Ludwig Mill. The first mill was constructed in 1822 by French Canadian trader Pierre Menard, later known as Peter Manor. Manor had been in the area about two decades when he built the mill on land granted to him by the Ottawa Indians. The mill, first used for sawing lumber, was used as a gristmill several years later.

Unfortunately, during the early 1830s, the proposed route for the Wabash & Erie and Miami & Erie Canal went through his thriving mill on the banks of the Maumee River. The State razed Manor's Mill but later gave him perpetual water rights to the new canal to power a new mill. The new mill construction began in 1847, but Manor died that same year and never saw the mill completed.

Isaac Ludwig bought the mill in 1865. Ludwig's first saw, a vertical saw, was replaced with a rotary saw. This is the type of saw still in use today at the mill. Ludwig ran the mill successfully until 1886, when he sold to Augustine Pilliod Jr., who renamed it the Pilliod Mill.

Frank Heising bought the mill from Pilliod, after Augustine returned from World War I in 1919. Heising died in 1931, and his son Cleo bought the mill. In 1971, Cleo Heising sold the business to Cleo Ludwig, a descendant of miller Isaac Ludwig. He then donated the mill to the Metroparks of the Toledo Area in honor of his great-grandfather. The mill then became the Isaac Ludwig Mill.

Isaac Ludwig Mill is an operational mill. Waterpowered turbines run the sawmill and the upstairs gristmill. The mill also houses a turn of the century machine shop, including a forge, power hammer, and other line shaft driven machinery, such as a wood lathe and drill press. Volunteer artisans work as tinsmiths, blacksmiths, spinners, and more. The mill is also occasionally run on steam. Providence Metropark is also home to a mule drawn canal boat. For more information call (419) 832-6004.

Story and photo courtesy of Steve Dey.

Isaac Ludwig Mill
Photo by Steve Dey

Mill at Lock Seventeen

Ohio

The Mill at Lock Seventeen, built in 1887 - 1889, opened for business in 1890. Located in Lock Seventeen, Ohio, a small community in scenic Tuscarawas County, it remains one of the area's historic landmarks. Local materials were used for construction, cut stone for the foundation and oak timbers for the three story post and beam structure. Some of the timbers in the mill are twelve feet and forty feet long!

Situated on the west side of the canal lock, it took advantage of the ten foot drop in the Ohio and Erie Canal elevation. Rail highway and waterway served the mill at its peak. A wide undershot wheel originally powered it. Over the years, the water works were destroyed and replaced with oil and then electric power. The flood of 1913 eliminated the canal. There is currently no lock or water at the site.

Put into operation as a flour mill in 1890, its output evolved to agricultural feed and supplies in the 1940s. It underwent a structural renovation in the early 1990s in which the mill building was restored and updated to meet state building codes. It still contains many belts, pulleys, and elevators from its early days. Today the mill houses commercial/retail stores. For more information call (740) 498-4278.

Story and photo courtesy Jim Russell.

The Mishler Weaving Mill

Ohio

The Mishler Weaving Mill, owned and operated by the Smithville Community Historical Society, provides a glimpse of a successful venture begun over one hundred years ago.

In the early 1880s, John Mischler brought his wife and young family to the area from Canton Bern, Switzerland with five dollars and a small trunk holding all their worldly possessions. After a time, the family settled in Smithville and began weaving rag rugs on a hand loom for local residents. The success of this business led to a more extensive weaving operation involving large power-driven looms. The product line included dish cloths, bath towels and washcloths, in addition to rugs. In later years, cheesecloth became a mainstay of the operation, supporting the large regional cheese-making industry.

Since purchasing the mill in 1993, the members of the Historical Society continue to operate the original Mishler looms. For more information call (330) 263-1481 or (330) 669-3231.

Story and photo courtesy of Nita Downey.
Photo by J. Irvin.

Ye Olde Mill
Ohio

The Olde Mill, built on this site in 1817, was one of the largest mills in the Ohio frontier. A second mill replaced the original mill in 1827, and the present mill was built after the Civil War.

An up and down sawmill provided cut lumber for construction of the early gristmill and both mills were built by James King, a well known Licking County millwright. In 1815, two years before building the first mill, he was one of Ohio's millwrights who agreed to a schedule of prices they would charge for their work. Again, in 1817, he met with a number of millwrights at Zimmerman's Tavern in Mt. Vernon to update the schedule. King, although considered eccentric, was well liked and respected for his ability.

The early mill had a waterwheel and millstones but not the refinements for milling found in those built two or three decades later. The impounded water would be conveyed through a millrace from above the dam to the forebay and penstock to turn the overshot wheel. The stone dam stood a short distance upstream. A flood changed the course of the river and destroyed the first dam and it was replaced with one built of brush and logs. Stonework of the original dam may still be seen but the brush dam washed away. The gristmill continued operations until the 1860s.

Although little is known about the second mill, there is reason to believe that it was not built until 1827. In January, Lemuel Jones executed a deed to John McNaughton covering a millrace at the site, and at the same time Eva P. Harris made a deed to McNaughton for mill privileges, "beginning at the southwest end of the dam." These documents suggest modification of the water supply for the mill, probably to meet the increased demand.

The second mill also operated as a waterwheel and buhrstone gristmill. The present, or third mill, built subsequent to the Civil War, was equipped in a like manner. Constructed of immense timbers, it had bigger grinding stones and the huge cog wheels, constructed of birch or applewood, transmitted the power from the waterwheel fed by a canal from the creek. The twelve by twelve beams used enabled the mill to withstand the shock of the operation of the waterwheel.

In 1866, J. Betts is identified as the owner of the mill property but thought to be a miller working for McNaughton. The mill, built about 1870, was done a floor at a time, one sitting on top of the other, with fourteen inch corner posts helping support and keep it in place. In addition to the waterwheel, the equipment consisted of three run of buhrs, bolting reels and other items needed to manufacture flour by the old-time method.

The roof was gray slate with the words "Cash for Wheat" outlined in black slate on the north side. The south side bore an outline of the date "1817".

Velvet Ice Cream Company, owned and operated by the Dager family for three generations, purchased the McNaughton Mill in 1960 and restored it over the next two decades. When fire destroyed that mill in 1986, the company built Ye Olde Mill to stand proudly on the same nineteenth Century hand cut stone foundation as its predecessor. The mill is now a restaurant, museum, gift shop and ice cream parlor. For more information call (740) 892-3921.

Story and photo courtesy of Mike Dager.

Butte Creek Mill

Oregon

The last original operating waterpowered gristmill in the Western United States is the Butte Creek Mill in Eagle Point, Oregon. Owned by miller Peter Crandall, it still produces nearly fifty whole grain products. This mill was built in 1872 and still uses the large four foot diameter French buhr mill stones to grind the grains the old fashioned way.

The sturdy old building, one of the first flour mills built in the Rogue River Valley, has stood on the banks of Little Butte Creek all of these years. The mill has a basement and a turbine where waterpower is produced and three floors where grain is received, stored, and ground. Architecturally the building is interesting because the frame was raised first. The large beams were mortised together and pinned with hard oak pegs called "tree nails." The walls of whipsawed lumber were nailed to the frame with square nails. Foundation pillars are two feet square and were hewn with a broad ax.

The entrance from the long wooden loading dock leads to the cool, shadowy interior of the mill. There lingers the tangy fragrances of wheat, rye, and corn, and the soft rhythmic sounds of rushing water turning the millstones fill the air.

The stones were quarried in France. Each weigh 1400 pounds. Shipped around the Horn to California, they were then carried over the mountains by wagon more than 127 years ago.

The lower level houses a facinating array of belts, shafts, and pulleys. Water diverted from the stream through the millrace activates the turbine turning the wheels that supply power to the mill. The water returns to the stream through the tailrace.

The mill is on the National Register of Historic Places. Peter Crandall spent years restoring the mill. Much of the machinery has been rebuilt.

The milling room and packaging and storage rooms have been renovated to meet current food inspection laws and assure proper sanitation of products.

This mill has a waterpowered refrigeration system that operates a freezer locker and cooler rooms. The 1938 Cyclops Ammonia Compressor housed in the basement is driven by this waterpower system.

The mill is currently operational and has a country store. For more information call (541) 826-3531

Story and photo courtesy of Peter Crandall.

Thompson's Mills

Oregon

Situated in the heart of Oregon's mid-Willamette Valley, Thompson's Mills is the sole surviving water-powered flour mill deriving its power from the Calapooia River in Linn County. Between 1847 and 1900 a total of seven flour mills were built on the Calapooia to meet needs of pioneer settlers and gold miners in California and southern Oregon. Boston Mills, as it was called at the time, was built by R.C. Finley, P.V. Crawford, and Alex Brandon on part of the Americus Savage donation land claim. Situated near the pack-train route, the mill intercepted freight traffic to the gold fields. A pack train might include as many as 150 mules carrying such necessities as flour, bacon, butter and whiskey. In this way, investment capital was secured to permit farmers and businesses in Oregon to develop and prosper.

The village of Boston grew up west of the mill and included a general store, blacksmith shop, harness shop, saloon and several residences. A nearby school doubled as a Methodist church on Sunday. In 1862, Boston Mills suffered a destructive fire, believed to have started in a wool carding mill. Because of the advantageous location, it was quickly rebuilt. The owner/ miller, William Simmons, served as postmaster for several years. The arrival of the Oregon and California Railroad resulted in the new town of Shedd and put an end to the village of Boston. The mill, however, continued to prosper.

Millstones were originally used, but in 1892 Martin Thompson, the new owner, replaced them with modern steel rolls. Architecture of the mill changed frequently to accommodate the need for more storage and machinery. But within the building shown in the photo, hand-hewn timber framing of the original structure remains. These changes ceased about 1930 as flour milling went to larger, more modern mills. Housewives stopped baking bread at home and local farmers shifted their crops from wheat to grass seed. Flour milling ceased in the early 1940s, but feed milling continued until the 1980s. The current owners converted one of the three existing turbines to generate electricity for the local utility company. The mill still generates power and does custom milling. The original three and one-half acre mill site has expanded to twenty acres.

In addition to the mill building, three historic support buildings and the mill owner's house survive. The Queen Anne style house was built in 1904 by Martin Thompson who gave the mill its present name.

Thompson's Mill is the oldest of two operating waterpowered mills remaining in Oregon. A significant historic and cultural site, Thompson's Mill should be cherished and preserved for the benefit of succeeding generations.

Story courtesy of Martin E. Thompson.
Photo courtesy Connie Petty.

Anderson's Mill

Pennsylvania

Nestled in the Cumberland Valley, is a mill complex built by Archibald Irwin in the 1760s. Included was a gristmill, sawmill, cooper shop, butcher shop, stone house, log house, post and beam constructed house, log barn, log hog pen, log smokehouse, a lime kiln, and other outbuildings.

Flour, cornmeal, meat, and other supplies were transported by as many as one hundred pack horses to supply the western army at Pittsburgh during the Revolutionary War.

Irwin built the home of limestone from nearby fields. A son, Archibald Irwin II, and his wife Mary succeeded to the old homestead. While visiting aunts in Ohio their daughters met the Harrison sons. Jane married William Henry Harrison, Jr. and became hostess of the White House for her father-in-law, President William Henry Harrison, in 1841. Elizabeth married John Scott Harrison and became the mother of Benjamin Harrison, the twenty-third President of the United States.

The mill, completely destroyed by fire in the 1840s, was rebuilt on the same site in 1856. It has a limestone foundation and post and beam construction, is approximately forty feet by sixty feet and has four floors. Originally there were four runs of French buhrstones, (two sets of stones were three feet in diameter and two sets four feet in diameter). The two flour buhrstones were replaced about the turn of the century with the first set of flour rolls to come into Franklin County. The last modification occurred in 1918, with the installation of an attrition mill for grinding feed. It replaced one set of the four foot stones, leaving one set of stones still operational in the mill. It has a full line of equipment including corn sheller, corn crackers, corn crusher, cockle machine, attrition mill, grain cleaner, flour rolls, flour dresser, wheat scour, flour packers, baggers, elevators, speed indicator, shafts, gearing (wood and metal), and belts. The mill is powered by two turbines, which replaced a waterwheel. The millpond has a one hundred twenty-five-foot long dam with an average height of ten feet.

The mill also contained a general store and office. A potbellied stove surrounded with boards about seven inches high, built in a square on the floor, sits in the store. The square was filled with ashes to catch hot coals that fell from the stove. It also provided a place for men to spit tobacco juice while arguing politics and discussing the problems of the day. Frequently, there were social gatherings at the mill, which led to lively conversations, music, and telling ghost stories. The room also contained a long wooden bench, a miller's desk, clock and related paraphernalia.

There is a small square trap door in the ceiling above the potbellied stove to let heat escape upstairs to another small room, now used as a shop to keep the mill in good repair. From 1856 to about the turn of the century, it served as an overnight room for travelers. Traces of the original wallpaper remain. The room had beds, a nightstand, possibly a candle stand and other items to make the travelers as comfortable as possible. There is no record of charges for an overnight stay but one unique event recorded is the birth of a baby.

The mill did business under the names Irwinton Mill, Union Mill, Union Roller Mills and Andersons Mill. Chester Anderson purchased the mill property in 1917 and it is still in the family.

Driving upstream along the millpond, the road follows a path with a picturesque setting with a peaceful meadow on one side of the millpond and limestone cliffs on the far side. For more information call (717) 328-3335.

Story courtesy of Harry Anderson.
Photo by Paul Nero.

Bahr's Mill

Pennsylvania

Bahr's Mill is a stone structure built to house Jacob Bahr's (pronounced Bear) woodworking and gristmill business. It is part of a farm complex in the village of Gabelsville, Berks County, outside Boyertown. The site had a linseed oil mill from ca. 1830 to the 1890s. Conversion to woodworking and production of wagon wheel spokes and tool handles took place in the early 1870s. The spoke-making machines, which include a rare "Blanchard" style duplicating lathe and a tenon cutter, have 1870 and 1871 patent dates. A non-automatic lathe used for wagon hubs and stair balusters is capable of turning pieces up to eight-feet long. The mill was rebuilt in 1897 and some time before 1945, Bahr's son, Lawrence, added machinery to make wooden hay rakes.

Gabelsville got its name from the Gabel family, who owned 200 acres there about the time of the American Revolution. This land included a gristmill on a different site dating to the 1750s. Abraham Gabel, born 1799 and his father Johannes, born 1766, were probably responsible for the old mill. It is also possible that Heinrich, the father of Johannes, who died at age 93 in 1828, was involved. Abraham and his son, John, later built the sawmill and bought the spoke-making machinery. John's daughter, Mary, was his only child to survive to adulthood. She married Jacob Bahr in 1888. The Bahr family has owned property in the area, according to their first appearance on the county tax records, since 1767.

The turbine-powered sawmill, built ca. 1860, operated until 1968 and provided raw material for the spoke factory. Both mills had a millrace connected by a tunnel under Sawmill Road. In 1945 the combined water sources were no longer adequate. A contract from the Boyertown Burial Casket Company for Bahr's sawmill required more hours of operation than the water provided. At that time, the woodworking business was largely ignored until Lawrence Bahr closed the sawmill in 1968.

At his "retirement", age 74, Lawrence returned to rake making, running the lineshaft with power from his tractor. He continued to work until his death in 1992 at age 98.

The modern history of Bahr's Mill includes election to the National Register of Historic Places in 1991, opening as a non-profit museum in May 1998 and delineation of the overshot waterwheel for the Historic American Engineering Record in the fall of 1998. Power is now supplied by an electric motor. The waterwheel will not be destroyed because of its rarity and historical significance. For more information call (610) 369-7266.

Story and photo courtesy Lisa Reber.

Brockerhoff Mill

Pennsylvania

The Brockerhoff Mill is a large four-story brick structure built about 1862 in the vicinity of Bellefonte, formerly known as Roopsburg. The exterior architecture features a handsome Federalist facade surmounted by an Italianate cornice and intersecting gable roof. The interior features massive post-and-beam construction, deep windowsills, and a first-floor atrium, with balcony, open to the limestone-walled basement area. There a small spring-fed stream falls into a sizable stonewalled pool stocked with trout. The pool is the former wheel pit, where the underwater beams that once supported two waterwheels to power the mill can be seen. One side of the pool has fine stone arches where water from the half-mile long headrace formerly entered. The spring-fed water (about 75,000 gallons per day) exits at the east end through an arched tunnel under the road and into the tailrace, where it eventually joins Spring Creek.

Henry Brockerhoff built the mill. Brockerhoff, a wealthy German and young man educated in France, served as an undersecretary to Napoleon. After Napoleon's banishment in 1815, Brockerhoff sailed for Australia, but his ship wrecked and the passengers were transferred to an American ship bound for Philadelphia. He eventually arrived in Bellefonte in 1825, a town of growing importance and wealth whose commercial buildings and residential mansions even today bespeak the town's nineteenth-century importance. As one of his many business ventures, Brockerhoff acquired the Roopsburg property in 1847 and had built the current mill by 1862.

Once the hub of a thriving village, the mill had four runs of stone and produced seventy barrels of flour a day. It ceased operation about 1940 and the village, sawmill, dance hall, and brewery complex had, to a great extent, disappeared. Today only the mill and the houses of the miller, brewmaster, gunsmith and blacksmith remain.

After WWII, a new owner sold the milling machinery for scrap, the granary and barn for lumber, the miller's house, and finally the mill, all in less than a year. In the ensuing decades the mill became a warehouse, dance hall, food market, church, and was finally abandoned. A local investor purchased the mill in the 1970s and undertook the renovation, stabilizing the structure and converting the first floor into living space. The current owner is continuing renovations.

The Brockerhoff Mill is listed in the National Register of Historic Places. It is currently a residence, artist's studio and occasionally a bed and breakfast. For more information call (814) 355-1003.

Story and photo courtesy Peter Schneeman.

Bromley Lowe Mill
Pennsylvania

Lumberville, PA
Photo by I. George Bilyk
Courtesy Society for the Preservation of Old Mills

Clear Springs Mill

Pennsylvania

The mill at Clear Springs (York County) first appears on the local tax records in 1809 when Adam Ernst built a gristmill and sawmill on 370 acres along the north branch of the Bermudian Creek. A generation later the mills, a still, and thirty acres of land were acquired by George Dick, and remained with that family until 1994. A village developed around the mills and eventually boasted a blacksmith shop, tannery, distillery and general store.

In 1886, business was booming and Edward Dick needed a larger mill. The wooden overshot wheel was protected inside the stone foundation. The simple two story pitched roof structure was replaced with three stories under a mansard roof and touches of country elegance in the Second Empire architectural style. The operation included three runs of stone in the gristmill and an additional run in the sawmill, used for grinding limestone. A shingle mill in an adjoining shed ran off a Frick Eclipse stationary engine mounted inside the gristmill.

In the 1920s the new generation replaced the old wooden wheel with a twenty foot Fitz overshot and steel forebay. A Leffel turbine had been installed and ran the flourmill, while the waterwheel provided power to grind animal feed and cornmeal with the other two runs of stone.

By the 1940s, business was no longer profitable. Most of the iron and steel sold to scrap mongers during and following World War II, Clear Springs Mill ground its last sack of cornmeal in 1947.

The mill became a giant catchall farm shed. A huge hole was cut in the back wall to move a hammermill inside. As time went on, the seams of the standing-seam metal roof loosened one by one, and the beautiful fish-scale slate on the mansard roof began to deteriorate. The dormers started leaking at the joints. Windowsills rotted away, while tree branches, pigeons and vandals took their toll on the sash.

In the 1950s the state constructed a new highway that separated the old mill from the water source. Meanwhile, poison ivy vines the size of an arm provided ladders for the raccoons to the third floor and ivy shoots forced their way between the wood siding. The dutch doors were broken by vandals and repeatedly boarded up, using larger nails each time.

Finally in the 1980s family descendants removed the millstones for mementos and held auction to sell most of the remaining equipment and miscellany accumulated in the mill.

Late in 1994 the gristmill, sawmill ruins and six acres of land were acquired by the Bert family, nearly fifty years after the end of operations.This marked the turn-around for the mill at Clear Springs. Out of love and reverence for a beautiful piece of history, the Berts launched a major restoration effort. The project, completed in the spring of 1996, received the 1996 Award of Excellence from the Pennsylvania Historic and Museum Commission for accurate attention to detail in restoration. The mill functions as a local living history site. The first floor has a gallery shop.

As of the date of this publication (1999) plans are to begin restoring the 1809 sawmill structure.

Clear Springs Mill is listed in the National Register of Historic Places. For more information call (717) 432-7702.

Story and photo courtesy Art Bert.

JOEL HILL SAWMILL

PENNSYLVANIA

The Hill Sawmill in Lookout was built by Holbert and Branning, shortly after the Civil War. In 1878, Joel Hill purchased it for $6,500. The purchase included 1,500 acres of timber and Equinunk Pond, currently known as Duck Harbor. The mill remained in operation until 1974.

The Hill Sawmill was powered by water. When winter arrived, the water froze and stopped the sawing operation. The time was put to good use by gathering timber for the next year's sawing. Loggers and farmers would bring in logs by horse drawn bobsleds and stacked them on the hill across the road from the sawmill pond. By spring, the whole field would be filled.

When the ice melted off Duck Harbor the sawmill employees would pull the log pile blocks (stop blocks) while two men flagged down traffic. The logs would roll down the bank, across the road, down the sawmill bank, and into the pond. The water from the pond helped clean mud and debris off the logs. This made it easier to cut the logs and did not dull the saw blade as quickly.

The mill could turn out 5,000 board feet of lumber a day, with their three-man crew. The turbine, turned by the force of water, was the power source for the entire sawmill. The turbine received energy from the force of the water running through the penstock, directed to the turbine.

Belts, ropes, and pulleys were engineered to handle the logs, as they were pulled into the mill and sawed into boards. The entire tree (log) was used. Nothing went to waste. The first slice off the log was used for firewood. The center became boards used in construction of homes and other buildings. Sawdust was hauled away by farmers to use as insulation for preservation of ice blocks (refrigeration) throughout the summer and as bedding for their farm animals.

Once a year, a lumber grader arrived to place a quality grade on each board cut. It took two to three weeks to grade the year's sawing. Local farmers then arrived with horses and wagons to haul the lumber to Kellam's Bridge. The boards, loaded onto logs lashed together to make a raft., were guided down the Delaware River to Philadelphia. The introduction of the train provided a different kind of transportation for the lumber.

The 120-year-old sawmill was given to the Equinunk Historical Society in 1988. It is currently listed in The National Register of Historic Sites and was a feature article in the Winter 1989 "Old Mill News" magazine. Since 1988, the Hill Sawmill Committee has been instrumental in working for the restoration of the mill. For more information call (570) 224-6722.

Story courtesy of Robert Kramer.
Photo by Stan Pratt.

HOCKLEY MILL

PENNSYLVANIA

The Hockley Mill, constructed about 1720, is located in Northern Chester County. The mill and the miller's house are currently owned by William and Camilla Anderson. Over the past ten years, William has done extensive restoration and both are listed on the National Register of Historic Places.

In 1718, Henry Hockley received a William Penn patent for 200.75 acres for twenty-four pounds and two shillings. He is thought to have been related to Richard Hockley, the business partner and attorney for Penn's son, John. Formal title was given in 1736 when the Lord Baltimore/William Penn line was settled. Occupation and use of the mill began around 1720. The gristmill ceased operation with the destruction of the millpond during a 1928 storm.

The sons of Henry Hockley were important figures in colonial Pennsylvania's preparations for the Revolutionary War. Thomas outfitted a battalion of infantry, serving as its colonel, at Valley Forge.

In September 1777, the Continental Army, George Washington and his Chief of Staff Alexander Hamilton, regrouped upstream from the Hockley Mill after the "Battle of the Clouds" at Malvern, Pennsylvania before moving on to the Battle of Germantown and then Valley Forge. With them was Lt. Michael Schall, the great-great-great-grandfather of William Anderson.

The Knauer family purchased the property in 1802 for about 1000 pounds, and, in 1805, rebuilt the log mill in stone as it is today. It was modeled from the new design of Oliver Evans in his 1795 book, *The Young Mill-Wright and Miller's Guide*, a very important work during the Industrial Age. The Knauers and descendants, with occasional interruptions, were residents of the property until 1965.

At its peak the millpond/race drew from the south branch of French Creek sufficent power to operate, although not all at once, a two-run gristmill, an up-and-down sawmill, a lath, drill press, grindstone, water pump, circular saw and DC generator. When the dam went out in 1928, the owners locked the gristmill door and allowed the mill remain in the 1805 condition. The sawmill sold in 1966 and was later reassembled in New Jersey.

Over the last ten years, the gristmill has been rehabilitated. It looks today, minus waterpower, nearly identical to the original mill, featuring parts handmade by the 1805 millwrights. In July 1997, the last missing gear was fabricated and installed. The waterwheel was turned by hand for the first time in almost seventy years. The mill now operates by recirculation of the aquifer water beneath the wheel. For more information call (610) 469-8636.

Story and photo by William B. Anderson.

Laughlin Mill
Pennsylvania

The earliest settlers along the Big Spring in the Cumberland Valley were the Scotch-Irish. These farmers and skilled workers were ardent Calvinists who left Scotland so they could practice their Presbyterian faith. They came to this part of Pennsylvania soon after 1730. These were educated men and women who purchased land, cleared it and built homes.

Among these earliest families were the Atchesons and Laughlins. In 1755 William Laughlin, a schoolteacher, married his pupil and neighbor, fourteen year old Mary Atcheson. Three years later, following the death of her father, John Atcheson, Mary inherited land with a mill along the Big Spring.

By 1763 William Laughlin had to pay taxes for the mill inherited by his wife. Laughlin's Mill had a very desirable location adjacent to the main road going through the valley to the west. It also had an abundant, never failing source of waterpower.

Although there were five other mills on Big Spring, the mill prospered and Laughlin added a stone addition on the east side of the log mill to accommodate the volume of business. Three generations of the Laughlins operated the mill, using millstones until the 1890s. The introduction of rollers in nearby mills rendered their mill unprofitable.

In 1896 the mill sold to the Newville Water Company, which used Cool Spring on the property as its source of water. The stone portion of the building was removed and the system's pumps installed in the old log mill and covered with weatherboarding.

In 1916 Dr. A. R. McCarthy purchased the water company. By the mid-1930s his daughter, Ethel McCarthy, who managed the company, had restored the log mill and added a screened porch on the south side of the building overlooking the dam and millpond. During the summers, she and her two sisters resided on the upper floor that had been converted into living quarters. The sisters lived there full time during World War II.

The Borough of Newville purchased the water company in 1953 and has owned the Laughlin Mill since. The chief plant operator of the Newville Water and Sewer Authority occupies the apartment and water pumps still occupy the lower level. The mill continues to be an important part of the community. Waters by the mill attract trout fishermen, as they have for more than 250 years, and the grassy park across from the mill is a favorite spot for photographers and painters. For more information call the Newville Historical Society (717) 776-6210.

Story courtesy Joan Brehm.
Photo by Steve Kennedy.

McConnell's Gristmill

Pennsylvania

In 1868, Daniel Kennedy rebuilt his original mill, previously destroyed by fire.

Thomas McConnell bought the mill in 1875 and improved it by replacing the waterwheel with water turbines and the grinding stones with rolling mills.

One of the first rolling mills in the country, it processed corn, oats, wheat and buckwheat for the community. In 1928 sagging profits and advanced technology forced the closure of the mill.

Thomas H. Hartman conveyed the mill and surrounding property to the Western Pennsylvania Conservancy and later to the Commonwealth with the wish that it be preserved for future generations.

McConnell's Mill State Park was dedicated in October 1957 and became a National Landmark in May 1974. The striking scenery and deep gorge is attributable to the glaciation of the area several tens of thousands of years ago.

Courtesy Society for the Preservation of Old Mills.

The Red Mill
Pennsylvania

Old mills have a fascination all their own and one of the oldest, built in 1805 is the Red Mill, near Buckhorn in Hemlock Township.

Elisha Barton, a native Virginian, born June 21, 1742, purchased 375 acres of land from Evan Owen, founder of Berwick. The land Barton bought, known as "The Elbow", extended lengthwise from the mouth of Fishing Creek toward Buckhorn and along Hemlock Creek.

The Barton family lived in their covered wagon until the father could erect a cabin. He then built a long barn and the frame mill. Barton installed a wheat stone chopper and plaster grinders and did custom work for farmers for many years. He owned and cleared land on the other side of Hemlock Creek and did considerable farming.

Barton gained prominence and for a number of years was justice of the peace. Part of the large tract he owned was later found to contain valuable iron ore and his son, Caleb, became quite wealthy from the proceeds. There were twelve children in the family.

Water from Hemlock Creek powered the mill. A headrace of earth and stone construction carried the water to an overshot waterwheel. The wheel measured approximately ten feet wide and one hundred feet in circumference. Tailrace or discharge water left the mill on the south side and re-entered Hemlock Creek.

The tax assessment of 1814 showed Barton paid one hundred forty dollars for seventy-five acres of land.

In March 1813 he sold the Red Mill to Lawrence Minegar and in 1828 Minegar sold the property to Peter Phillip Appleman and others. Under this ownership Noah Crites and George Dreisbach II operated it. Hemlock property taxes for 1828 totaled fifty dollars.

Isaiah William McKelvy, owner in 1882, remodeled the building and added a steam boiler at the west end. He also installed modern machinery for roller processing of wheat into flour. The Red Mill is now a furniture and antique store. For further information call (570) 784-7146.

Story and photo courtesy of Jay & Lenore Fritz.

Shiffer's Mill
Pennsylvania

Shiffer's Mill is a 1799 wood frame structure on a small farm midway between Elizabethtown and Hershey in rural Conewago Township. It was originally named Goss Mill, after the family who first built and operated it. Operations are thought to have ended in the 1920s while the Shiffers owned it. The wheel and machinery were removed and the mill was subsequently used as a barn for raising ponies a few years. Bob and Joyce Carbaugh purchased the structure in 1976. Pony stalls remain in the lower level.

The mill is identified in an 1883 book entitled *History of the Counties of Dauphin and Lebanon in the Commonwealth of Pennsylvania,* by William Egle, M.D. Page 422 of the book confirms the construction of three mills in the township, including Goss Mill built in 1799 as both a grist and sawmill. Prior to this time, the book states that farmers "hauled all their grain to Philadelphia over poor roads with little or no accommodation "for man nor beast" along the way. The teams at night halted by some stream of water, the feeding trough was fixed upon the wagon-tongue, and there the horses ate and slept, no matter how inclement the weather, the drivers stowing themselves snugly under the wagon cover in the "fuhrmons bet".

Philadelphia is about 150 miles east, so undoubtedly the construction of mills in the township provided a tremendous benefit for local farmers. A small creek with a dam located about one hundred yards upstream provided the water to power the mill. Water accumulated at night, flowing into a headrace when the gate opened. Approximately a half day of operation could be done before the waters depleted. It is believed the dam washed out during a 1940s flood. The races are now backfilled with little evidence remaining, except for a two foot diameter pipe under the adjacent road where the tailrace re-entered the creek. An older gentleman still living nearby worked in the mill as a teenager. He recalls ice skating on the creek about a mile downstream when a few inches of water would suddenly sweep over the ice as the races became active.

In 1994, an elderly couple stopped along the road as Bob Carbaugh painted a fence. The lady announced that she was born in the millhouse in 1900, moved to a nearby farm at age three, and now lived about an hour away in Coatsville. Her maiden name was Mengel, which coincided with the name "E. Mengel" stenciled on the inside of one of the mill doors.

When she saw the stenciling, her father's name, it looked as though she experienced a kind of spiritual event. She is now living nearby with her son, and has become a friend and frequent visitor.

After completing an ongoing restoration of the log house, there are plans to install a large "hit or miss" engine and thirty inch circular saw on the main floor of the mill for stove wood cutting. Other related antique equipment will also be placed and the original atmosphere created as much as possible.

Story and photos courtesy of Bob & Joyce Carbaugh.

Shoaff's Mill
Pennsylvania

Not so long ago a gristmill was as necessary to a farm village as a gas station is today. Hundreds of mills once dotted the landscape. In fact, many of the roads in this country were built over well worn paths to mills.

Imagine traveling part of an old highway to bring your wagonload of grain to the mill. When nearing the mill, the rumble of the giant waterwheel can be heard. Perhaps a miller's helper, dusty from his work, is outside the mill to help unload the corn. Your thoughts turn to taking home fresh-ground cornmeal for those tasty cornmeal muffins.

Flour milling became another of the original industries in Pennsylvania. However, ironmaking is considered responsible for the formation of Shoaff's Mill. In the early 1800s a self sufficient community grew up around Juniata Furnace. The first mill, located west of here, ground grain from company owned farms.

Shoaff's Mill, built about 1840, was part of the furnace complex. By this time the community consisted of the ironmaster's mansion, eleven tenant houses, a store and a blacksmith shop. Because of financial difficulties, the furnace ceased operations soon after the building of the mill. Several years later, William Shoaff purchased the mill and improved the operation. Shoaff expanded the business to grinding grain for markets, stores and local farmers. He also built the brick house across from the mill. After Shoaff's death in 1888, his wife continued operations.

In 1900 Ellis Shoaff purchased the mill from his mother at public sale. He operated the mill at full capacity, grinding wheat, buckwheat, cornmeal and animal feed. He replaced the wooden wheel with the present thirty-two foot steel wheel from Hanover. This waterwheel was used until the mill closed in the 1940s. It is rumored to have been one of the largest waterwheels east of the Mississippi.

Small quantities of clean, roasted corn were usually delivered to the mill by the owners of the grain. It would be dumped into the hopper by the miller, ground on the millstones, sifted and returned to the owner. The miller kept a portion of the meal, called the "toll," as payment for his services.

Long ago, this mill ground wheat flours and cornmeal. Grinding wheat flour, more complicated than cornmeal, required several additional steps. These included hoisting the wheat to the attic floor for cleaning, elevating the grain back to the attic for cooling after grinding, and sifting the ground wheat to separate the flour, middlings and bran. Future restorations may make it possible for Shoaff's Mill to demonstrate the grinding of corn for cornmeal and the wheat flour process. For more information call (717) 567-9255.

Story and photo courtesy Bob Cross.

Boyd's Eight-Vaned Wind Gristmill

Rhode Island

In 1810 John Peterson, a retired sea captain, brought from Maine two men, Brayton and Babcock, who built a mill in Portsmouth. It was a basic mill without sophistication, typical of the wind gristmills of Aquidneck Island. The oak timbers were cut near Wickford Village in North Kingstown and floated across Narragansett Bay to Portsmouth. Peterson also used oak knees from his wrecked schooner in the construction of the mill.

William Boyd ran a wind gristmill near Bristol Ferry that was destroyed in a storm in 1815. William then leased and eventually bought John Peterson's mill and rebuilt it on the present site. Succeeding generations of Boyds operated the mill, including Leander, Benjamin F. C., Clifton, Frank and probably the most well known operator, Benjamin Franklin Cornell Boyd (1861-1946).

Boyd's mill originally ground feed products for animal stock and family meal. For years feed grinding remained the principal business, and from 1840 to 1884 grains from almost every Aquidneck Island farm passed through the mill.

With money scarce, the miller got his pay by taking toll. Later regulated by law, a windmill operator took one-tenth of every bushel, or about three quarts. This custom came to an end about 1910.

With the development of the West's cheap grain and meat products, it became unprofitable for Rhode Island farmers to continue the old methods of farming. They went more and more into truck farming instead of raising grain and animals. This change marked the beginning of the end for many of the old windmills. Gradually they were abandoned and allowed to decay. Rhode Island corn possessed merits for making a fine, soft meal often used for making the Johnny Cake.

Realizing the grinding of feed to be a vanishing industry, the Boyd family pushed the meal business. Meal was shipped as far away as South America. The mill ceased operation in 1946. With its demise an important symbol of Middletown's heritage slipped almost unnoticed into oblivion.

While it may not be possible or desirable to try to preserve every mill, there are sound reasons for ensuring the survival, relocation and restoration of the Boyd's Wind Gristmill. Posterity is entitled to a share of the heritage we may have either ignored or enjoyed. For more information call (401) 849-1870.

Story and photos courtesy Stanley Grossman, President, Middletown Historical Society.

Carpenter's Gristmill

Rhode Island

The quaint little gristmill on Moonstone Beach Road in South Kingstown was built by Samuel E. Perry in 1703. It has been known as Carpenter's Gristmill since 1874 when the Carpenter family first became owners.

It is a working mill where Rhode Island Whitecap Flint Corn is stone ground by waterpower. In continuous operation since built, it is the only waterpowered mill still operating in Rhode Island.

The water supply is from Perry's Millpond lying easterly from the mill. A deed dated May 3, 1702 refers to the purchase of two and one half acres by Samuel E. Perry, to "drown a swamp by damming the brook, to make a pond to set up a mill."

The Perry family operated the mill until 1789. Over the years, it changed hands many times. While in the ownership of the Watson family, around 1825, the mill was taken apart and moved to the present location.

In 1874, Wanton R. Carpenter purchased the mill. In 1923, Stanton Leroy Carpenter acquired it. Heirs of Carpenter sold the mill property to Mr. and Mrs. Rowland Robinson in 1964. Robert and Diane Smith purchased the mill in 1986. Restoration began in 1988. Carpenter's Mill was placed on the National Register of Historic Places in 1990. For more information call (401) 783-5483.

Story and photo courtesy of Diane & Bob Smith.

EXETER HOLLOW MILL

RHODE ISLAND

The Exeter Hollow Mill is in the town of Exeter approximately twenty miles south of Providence.

In 1975, the mill site, consisting of approximately three acres and a pond, was purchased. Eight interesting and extremely busy years were spent researching and building a new "old mill" on the Queens River property.

It is not known when the original mill was built, but in 1847 the "factory", sixty-eight feet long by twenty-eight feet wide, three stories and an attic high, leased as a cotton factory with waterwheel.

The Hendricks designed and constructed the present mill as near to the research description as possible. With the help of a back-hoe, large bulldozer, couple of old trucks, two healthy sons and much ambition, many hours were spent bulldozing the original six to seven acre pond, ridding it of thousands of yards of mud, fill, and boulders, and rebuilding the tumbled down spillway.

The five hundred foot headrace had to be cleared. The retaining walls at the end of the headrace were caved in and the patched, rusted out thirty foot flume dropped twenty-four feet to a broken turbine. All of this was removed and rebuilt. At present a turbine is in the mill.

A first class stonemason was employed to work on all phases of reconstruction and new construction. The spillway and waterwheel pit of stone were rebuilt. The original waterwheel, a thirty foot diameter breast wheel, generated the power for a cotton mill. Also on the property (1847-1869), sat a gristmill, sawmill, blacksmith shop, four tenement houses and a store. The stone tailrace ran underground for approximately 100 feet. It had partially caved in and was filled with junk.

It proved exciting to build the mill entirely of post and beam construction. The existing mess on the original building site was removed and work commenced on the foundation and stone construction of the ground level of the three story mill.

With the knowledge and help of a builder known as the "house smith" and the mason, the posts and beams started to take shape.

Large beams were purchased from Fall River, Massachussetts from old turn of the century mills. The siding was locally cut, native pine.

Cedar trees were cut from a local swamp, and the Hendricks made the cedar shingles with an old lane shingle mill.

The Fitz twenty-two foot by five foot overshot waterwheel was purchased in Pennsylvania from the Biesecher Family, who had advertised it in the *Old Mill News*. They helped dismantle the wheel from a difficult position within their mill (dog food mill). The operation took about a week. New buckets had to be fabricated.

Through the years, work progressed. The necessary cables, shafting, and pulleys procured, and the mill began generating electricity with a twenty-five kw generator, to supply the mill, greenhouse and home. Excess power is sold to the area power company.

A fifteen inch enclosed flume turbine was installed and a blacksmith forge built on the ground floor of the mill. The original blacksmith enjoyed his own separate stone building, but only the tumbled foundation remained.

The first floor, or basement, is where the blacksmith, tools and electricity are located. The second floor is the gristmill, which grinds locally grown corn into meal. The third floor is a woodworking and carpentry shop. *Old Mill News* January 1985.

Story by Paul Hendrick,
submitted in loving memory by his wife, Betty.
Photo by Herman Marshall.

SLATER MILL
RHODE ISLAND

Slater Mill Historic Site sits on five and one half acres of property with the Blackstone River running through it. It was the location of intense industrialization and technological innovation during the late eighteenth and early nineteenth centuries and has been referred to as the birthplace of American industry. Here, Samuel Slater successfully duplicated British waterpowered spinning, and fiber processing machines at Carpenter's clothier's shop in 1790, thus earning himself a partnership in the first waterpowered cotton textile factory in America. The yellow clapboard mill building was constructed in 1793. The management structure, architectural style and workforce were emulated throughout southern New England.

The rubble stone mill, built for Oziel Wilkinson in 1810, housed a machine shop and a cotton spinning mill. David Wilkinson operated the machine shop from 1810 to 1829 and developed the screw cunning lathe, an improvement of William Gilmore's powerloom, the first commercially successful powerloom in Rhode Island.

The 1758 Sylvanus Brown House is also preserved at this site. Brown was a master craftsman who made machine patterns and wooden machine parts for Slater' s textile mill.

The Old Slater Mill Association, formed in 1921 to rescue the mill. Restoration was completed by 1925. In 1966, the wooden textile mill earned designation as a National Historic Landmark. In 1971 the Wilkinson Mill came to the site. Restoration completed in 1973, it became recognized as a National Historic Mechanical Engineering Landmark by 1975. A waterpower system operates machine tools in an exhibit opened in the Wilkinson Mill in 1981. The Sylvanus Brown House, moved to the site in 1963 and restored, opened as an artisan's dwelling in 1975.

Slater Mill Historic Site is now a museum complex dedicated to the preservation and interpretation of American industrial heritage. For more information call (401) 725-8638.

Story and photo courtesy of Slater Mill Historic Site, Pawtucket, Rhode Island.

DEKALB COTTON MILL

SOUTH CAROLINA

The Kendall Wateree Plant in Camden was built in 1899 as the DeKalb Cotton Mill by W.B. Whalley. A classic example of late nineteenth century architecture, it was one of the show places of this area. The mill village consisted of about seventy homes. During the first seventeen years, several different firms owned the plant and added more machinery and village houses.

In 1916 H.P. Kendall acquired the mill. The first southern acquisition, it became the cornerstone of the Kendall Company Weaving Division. Kendall modernized the plant in 1921 by replacing the old plain looms with 420 new E model Draper Automatics. This improvement was followed by other vital improvements, such as hydroelectric power, new humidifying system, new lighting, additional card room machinery and the building of twelve new houses in the village. In addition, the mill and village homes were repainted. In 1948 these homes were sold to the employees dwelling in them. Although the historic beauty has been preserved, the Kendall Wateree Plant has undergone several major modernizations in equipment and operations since 1921.

The Wateree Plant has a remarkable record of continuous operations since 1916. During World War II, the 200 employees at this plant worked faithfully on a six-day schedule to keep production going to produce gauze material. In 1972, the Kendall Company became a wholly owned subsidiary of the Colgate-Palmolive Company, a world leader in the production of consumer products. In 1983 shuttle looms were removed and air jet looms, new opening, carding, spinning and cloth inspection equipment introduced.

In 1988, the Kendall Company became a major contributor to the newly created Healthcare Products Division. In 1994 The Kendall Company was acquired by Tyco International Ltd. Inc. and is now a part of Tyco Healthcare. In 1998 a major modernization took place with the addition of new state of the art open end spinning equipment.

Story and photo courtesy Sandra Brown.

Ellerbe's Mill
South Carolina

Ellerbe's Mill is in Sumter County on Rafting Creek, which flows into the Wateree River. At one time there were at least four watermills on Rafting Creek. Ellerbe's Mill is the only of its kind remaining.

Garner Sanders built the original structure about 1835, on land inherited from his father, William Sanders. The millpond, built with slave labor, is approximately ninety acres. The son of Garner Sanders, William Marion Sanders, died at an early age. Executors of the estate sold the mill, pond, and surrounding plantation to a cousin, William Crawford Sanders Ellerbe, in 1861. The mill was remodeled about 1880 and a cotton gin and sawmill were added. All were powered by water pressure, but the additions no longer exist.

In April 1865 General Edward E. Potter of the Union Army made what is now known as "Potter's Raid", from Georgetown. He burned and destroyed railroads, cotton, mills and homes. When a detachment of this army passed Ellerbe's Mill, a faithful slave held the owner's Masonic apron before the millhouse door. The commander of the unit, also a Mason, spared the mill.

Between 1933 and 1970, Perry M. Brown operated the mill. The plantation, mill and pond are now the property of James L. Haynsworth, great-grandson of William C. S. Ellerbe and nephew of Perry M. Brown.

Ellerbe's Mill is listed on the National Register of Historic Places. Although not operational for fifteen years, the owner hopes to restore it to operational status.

Story and photo courtesy James L. Haynsworth.

Golden Creek Mill

South Carolina

Golden Creek Mill in Early is a wonderful example of nineteenth century mills that provided flour and meal for the early American family.

This mill was first built across the creek, circa 1815, where the ruins are still visible. Joseph Woodal (O'Dell) received the land grant on Golden Creek June 5, 1786. The family (originally spelled Woodhull and pronounced "Wodle") came from Northamptonshire and Oxfordshire, England. William married Martha Mullinax and they had three sons, John Calvin, Frederick Garvin and Perry O'Dell. William O'Dell was granted land to operate a gristmill. Materials for the impressive old dam were constructed on the premises.

By 1835 the area was becoming a cotton center, and O'Dell sold the property to John Arial, who added a cotton gin and press. Other owners of the mill site were John and Alfred Arial (1836 - 1886); Hagood and Alexander (1886 - 1891); A. M. Mauldin (1891 - 1899); D. M. Mauldin (1899 - 1903); and George H. Hendricks (1903 - 1985).

In 1985 Leroy Stewart purchased the land from the Hendricks Estate and began construction of the gristmill. A stone foundation ten feet high and eighteen inches thick at the base elevated the two story frame structure above the flood level of the creek. Parts for the mill were salvaged and assembled from the states of Oregon, Minnesota, Nebraska, North and South Carolina and Michigan. The millstones were purchased in Copenhagen, Denmark and shipped to Charleston, South Carolina in 1987. The millrace was built by Leroy Stewart with the help of his uncle, Furman Stewart. They also constructed the first waterwheel of wood, but later replaced it with a fourteen foot metal wheel built by Bob Johnson and son, Bobby, assisted by Stewart. When moved to the present site in 1990, the mill began grinding corn into cornmeal and grits.

The second floor of the mill is a gift shop and Cliff Grossenbaugh, who handcarves and paints birds native to the area, is usually found there. For more information call (864) 859-1958.

Story and photo courtesy of Leroy Stewart

Hagood Mill
South Carolina

Pickens (Pickens County), SC
Photo by J.C. Barger

Horton's Mill
South Carolina

Chesterfield County, SC
Photo by J.C. Barger

Preacher Raines Mill
South Carolina

Westminster (Oconee County), SC
Photo by J.C. Barger

Prices Mill
South Carolina

Parksville (McCormick County), SC
Photo by J.C. Barger

Silas Butts Mills
South Carolina

Westminster (Oconee County), SC
Photo by J.C. Barger

The Dakota Flour Mill

South Dakota

The Dakota Mill in rural South Dakota is a restored roller mill. Ray Osman of Redfield, South Dakota bought it in 1990 from Dewey Sheetz of Muncie, Indiana. The mill has been traced to Enon Roller Mill of Enon, Fast Bend, North Carolina but had been out of use for many years. The mill was built by XX Century Mill Manufacturing Co. of Decatur, Illinois in the early 1920s. It bears the serial number 153 and is a twenty-five barrel roller mill with four sets of rolls; two sets of corrugated and two sets of smooth. After each set of rolls a reel covered with cloth sifts the flour. Finally it falls to the bottom of the mill where it is taken by elevator to a holding bin.

Mr. Osman brought the mill to South Dakota from North Carolina on a large flatbed truck trailer, causing quite a commotion as it traveled down the gravel roads to his farm. He spent the next several years dismantling it and sandblasting the rust. He cleaned it, repaired it, put sealed ball bearings in place of the cast iron bearings and replaced the cloth and brushes on the rollers. After erecting a building to house the mill, he built a network of bins and feeders to supply the mill with five bushels of wheat per hour. This is equivalent to about one barrel or 196 pounds of white flour per hour. An elevator was also installed to carry the flour to a bin once it has been ground.

Finally, in 1995 it was ready for use once again. Mr. Osman operates the mill regularly for his own use and to share the wheat he raises with his neighbors. He has been farming in the area for over sixty-five years cultivating some of the same land his parents did when they homesteaded over one hundred years ago. The wheat used is mostly hard red spring wheat that weighs approximately fifty-eight to sixty pounds per bushel with thirteen to fifteen percent protein. A tempering period is required to soften the shell so the milling process takes several days. It is then sifted with a Meadows Sifter and bagged. In addition to wheat, Mr. Osman also grinds whole wheat rye, buckwheat and oat flour with a stone burr mill and an attrition mill. For more information call (605) 472-1978

Story and photo courtesy Ray Osman.

Bell's Mill
Tennessee

Ooltewah (Hamilton County), TN
Photo by J.C. Barger

Blowing Cave Mill
Tennessee

Bird Crossroads (Sevier County), TN
Photo by J.C. Barger

Bush Creek Mill

Tennessee

West Point Community (Lawrence County), TN
Photo by J.C. Barger

Davey Crockett Mill
Tennessee

Lawrenceburg (Lawrence County), TN
Photo by J.C. Barger

John P. Cable Mill

Tennessee

John P. Cable Mill is located in Cades Cove. There were hundreds of gristmills in the area now included in the National Park. Most were small family mills but a few were large community mills like the John P. Cable Mill. This type of mill usually operated under a toll system. The miller would receive a gallon of grain per bushel for grinding it. The mill was originally built by John P. Cable and has been restored twice.

Photo by Walter Cline.
Courtesy Society for the Preservation of Old Mills.

John P. Cable Mill
Smoky Mountain National Park (Blount County), TN
Photo by J.C. Barger

The Crosseyed Cricket

Tennessee

Since the restoration of this gristmill in 1976, it has become a centerpiece of the Crosseyed Cricket Resort. It was built by Gideon Capps in the 1850s. His son, Mark Capps first recorded it on a Civil War Direct Tax Assessment dated June 7, 1862.

Between 1876 and 1879 a sawmill was added to the existing mill by James A. Cardwell and operated approximately twenty-two years. When A. C. Moore purchased the mill and three surrounding acres in 1923, he discontinued use of the sawmill. He, his wife Ida, and their three sons lived on the second floor of the gristmill for two years while he used the waterpowered sawmill to cut timber for a home for his family.

An overshot wheel was powered by water from a three-acre millpond fed by Big Blue Spring. In the 1920s the millpond washed out resulting in a half-acre millpond. The smaller pond never had the water reserve needed and, although smaller stones were substituted, the mill never operated economically after that time.

When Elmer Hudgens bought the property in 1929, he renovated the mill and hired a miller to run it. He also built a two-room log cabin which now houses dining rooms for the Crosseyed Cricket Restaurant.

In 1975 Jim Lockwood began restoration and changed the water flow, converting to a breast wheel. The water from Big Blue Spring now flows through The Crosseyed Cricket Trout Hatchery, half-acre Trout Lake and three-acre Catfish Lake before driving the wheel. Since its restoration in 1976, Mr. Lockwood has rebuilt the wheel a second time and replaced the cast-iron gears.

The exterior of the mill is true to the original and the interior is paneled with Poplar board and batten. The original meal-bin, hopper and stone-jack, as well as the smaller stones are used.

Story and photo courtesy of Jean Lockwood.

Evin's Mill

Tennessee

Smithville (De Kalb County), TN
Photo by J.C. Barger

FALLS MILL

TENNESSEE

Falls Mill is a three story brick and post-and-beam structure built in 1873 as a cotton and woolen factory. The site was surveyed as a "mill seat" in 1808. Not until the early 1840s was the natural fall in the creek harnessed to power a textile factory. By the 1850s this enterprise became known as Hunt's Factory. During the Civil War, the Criddle family owned and operated the factory, assisted by Robert Newton Mann.

In 1864 Mann partnered with Azariah David, whose textile factory had been burned by Union troops. Mann and David contracted to build the present mill in 1860. The current stone dam replaced a wooden dam in 1871. Waterpower was used for a sawmill to cut the timbers for the new factory and cotton gin building completed in 1873. Cotton was ginned, carded, and spun into thread and yarn, and wool was carded for home spinning. Young women and children were employed as operatives about eight months a year, working seventy-two hours per week for a two dollar weekly wage. Men supervised the manufacturing and earned about eight dollars per week. In 1876 Mann and David built a waterpowered flour mill further up the creek. The older factory building was dismantled in 1881.

Financial difficulties dissolved the Mann and David partnership in 1883. The property was acquired by the Lucas family. Installation of a new overshot waterwheel took place in 1886 and the operation continued sporadically until about 1906. Installation of the present thirty-two foot Fitz steel waterwheel and relocation of the millrace allowed conversion to a cotton gin.

In 1930, because of the advanced age of the owner and effects of the Great Depression the mill property was divided and auctioned. J.H. Walls continued operation of the gin until World War II. Following the war, Harry Dalton and his father bought the mill and converted it to a woodworking factory. The textile machinery was scrapped but the waterwheel continued to operate. In 1968, Colonel and Mrs. Woodrow Crum restored the mill and moved most of the present grain milling equipment from the Boiling Fork Mill in Winchester. The Crums sold to Donald Cunningham in 1974, and he sold to John and Jane Lovett in 1984.

The mill is listed in the National Register of Historic Places and sits within a National Historic District on Factory Creek in Franklin County. It is a fully operational grain mill powered by the thirty-two foot overshot waterwheel. Falls Mill now features antique grain and textile machinery, a country store, and log cabin bed and breakfast accommodations. It is also home to the Museum of Power and Industry, Inc., a non-profit educational organization founded by the Lovetts in 1981. For more information call (931) 469-7161.

Story and photos courtesy of John & Jane Lovett.

Falls Mill
(Riverdale (Franklin County), TN
Photo by J.C. Barger

Hord Mill
Tennessee

Summer Shade (Hawkins County), TN
Photo by J.C. Barger

Hyder's Mill
Tennessee

Bethel Community (Carter County), TN
Photo by J.C. Barger

King's Old Mill

Tennessee

King's Old Mill is located in Union Hill Community, Overton County, Tennessee. The first mill was built here in 1820. Restoration started in 1994 using period logs. A 20-foot Fitz overshot wheel was installed. Mill restoration is near completion in 1999. Owners are Kenneth & Anne King and Harvey and Helen Cummings.

Photo by Steve L. Spring.

Union Hill Mill
(Now known as "King's Old Mill")
Clarksville Vicinity (Overton County), TN

KETNER'S MILL

TENNESSEE

David Ketner, the family's first arrival in the Sequatchie Valley, made his living in a mill on Looney's Creek where he operated the original Ketner's Mill. The mill is found in what is now known as Ketner's Cove, near the base of Suck Creek Mountain. "Great-grandfather" David Ketner began his sunrise-to-sunset work in the gristmill on this site. He also operated a blacksmith shop and, in 1840, added a wool carding machine.

David's son, Alexander, ("Pappy") bought a mill site in 1868. The new site, on the Old Sequatchie River, stands there still. The oldest part is the wooden mill dating back to 1824. The "newest" part is the brick building. The bricks were meticulously molded by hand of the clay from a nearby hillside. The scooped-out hillside still shows signs of the hard labor. Construction began soon after the land purchase, but was halted by the War Between the States. Two members of the Ketner family, Henry W. and Thomas D., served as Confederate soldiers.

An old family story tells of the birth of a sickly child who, during the war, needed a doctor. Since no Confederate doctor was available, Mrs. Ketner suggested appealing to the Union doctor with troops occupying the area. Her husband, Alexander, flatly refused, saying, "He's a Yankee; he will kill him." Mrs. Ketner answered that she hoped he was "a doctor first, and a Yankee second." Legend has it the "Yankee" doctor was summoned, and the baby saved.

After the war, construction on the brick portion of the mill was resumed and completed in 1882. In the early 1880s, after moving the wool carding machine to this "new' site, the mill was up and running again.

The mill has been used since to grind corn meal and wheat. Giant millstones crush the corn and the kernel is left in the meal, making it healthier. According to enthusiasts, it tastes better than store-bought meal. The immense weight of the millstones, coupled with the deep cut geometric pattern, slice the grain to produce a unique texture and flavor. "Music" emanates from the millstones as they skip over thousands of pounds of grain. Clyde Ketner, the last member of the family to run the mill, wrote in *Old Mill News* that "Only after months or years of grinding do a miller's ears become tuned to the various sounds of the millstones. To the veteran miller, the 'singing of the stones' becomes a beautiful tune. No one can tell the young miller how to adjust the (machinery) for grinding. The sound is the only way."

From "great-grandfather" David Ketner through five generations of loving care, Ketner's Mill stands now, almost exactly as it was in 1868. In 1977, it was listed on the National Register of Historic Places.

Until the death of Clyde Ketner in 1992, this was one of the last remaining gristmills in the country to operate daily. Now the mill is operated only at fair time. For more information call (423) 821-3238.

Story and photo courtesy Sally McDonald.

Leslie Mill
Tennessee

Kingsport (Sullivan County), TN
Photo by J.C. Barger

Leslie Mill
Tennessee

Kingsport (Sullivan County), TN
Photo by J.C. Barger

Leslie Mill
Kingsport (Sullivan County), TN
Photo by J.C. Barger

Lindsay Mill
Tennessee

Scotts (Sevier County), TN
Photo by J.C. Barger

Norton Mill
Tennessee

Maynardsville (Union County) TN
Photo by J.C. Barger

Bud Ogle Tub Mill

Tennessee

Statesville (Sevier County), TN
Photo by J.C. Barger

Bud Ogle Tub Mill
Statesville (Sevier County), TN
Photo by J.C. Barger

Pigeon Forge Mill

Tennessee

The old mill, on the east bank of the west prong of the Little Pigeon River in Pigeon Forge was built in 1830. Water from the river turns the twenty-four foot breast wheel and two old tub wheels to furnish power.

Kept in continuous operation since built, the Old Mill is completely waterpowered, using the same aged machinery. It still grinds on old French buhrs (made from flint granite and weighing over two tons).

The mill is supported on fourteen inch by fourteen inch yellow poplar logs forty feet long, resting on pillars built of huge river stones. The interior is hand-hewn hemlock and oak held together with hickory pegs. The floors are heart pine put down with rectangular iron nails. The outside is yellow poplar weather boarding.

During the War Between the States looms were kept on the second floor and weaving done for the soldiers. The old iron forge, from which the town received the name, was a thriving business. The mill was built beside it and both used waterpower. In 1885 the forge was dismantled and taken to Kentucky and a waterpowered sawmill built. In 1900 the sawmill was moved and a large storage room for grain added to the mill. Pigeon Forge Mill generated electricity for the town until 1930.

For more information call (923) 453-4628.

Story and photo courtesy
Society for the Preservation of Old Mills.
Photo by Ernie Tucker.

Reagan Mill

Tennessee

Gatlinburg (Sevier County), TN
Photo by J.C. Barger

Rice Mill

Tennessee

Norris (Anderson County), TN
Photo by J.C. Barger

RIVERDALE MILL
TENNESSEE

Riverdale (Knox County), TN
Photo by J.C. Barger

Silver Lake Mill
Tennessee

Mountain City (Johnson County), Tennessee
Photo by J.C. Barger

Walker Mill
Tennessee

Westminster (Greene County) , TN
Photo by J.C. Barger

Alvin C. York Gristmill

Tennessee

Parksville (Fentress County), TN
Photo by J.C. Barger

Alvin C. York Gristmill
Parksville (Fentress County), TN
Photo by J.C. Barger

Barnard's Mill
Texas

Construction of the three story Barnard's Mill on the waters of the Paluxy River in Glen Rose was completed in 1860. The stone walls are three feet thick with a foundation twenty-five feet to bedrock.

Charles Barnard received 160 acres of land provided he build a gristmill to encourage settlement in the area. Gun ports used for protection against the Indians are located at the top of the mill (see photograph above). Gun ports are also in the shutters of the windows.

Prior to construction of the mill, Barnard, his brother George, the Torey Brothers and, silent partner General Sam Houston operated Indian trading posts along the Brazos River.

An outside waterwheel never existed. The millstones were powered by a turbo jet waterwheel in the basement. This was considered most innovative at that time. The mill operated on waterpower until converted to wood burning steam engines in 1890 and the two-story cotton gin was constructed.

Barnard's Mill served as the center of social and professional activities for many years. In 1868 the Glen Rose Masonic Lodge organized on the third floor. Court was held there as the community justice system. Settlers brought corn and grain to be ground. Overnight camping on the grounds and Saturday night dances were not uncommon. Bands came from as far as Waco, Texas to play at the mill.

In 1942 the mill was converted to a hospital for nearly thirty years until a new hospital was built.

Abandoned for several years, Richard H. Moore purchased the mill in 1979. It is now his home after nearly seven years of restoration and furnished with period pieces from 1775 to 1875. Mr. Moore loves tell the many interesting stories surrounding the history of the old mill.

Barnard's Mill has been on the Texas Registry of Historical Places for over thirty-five years and, in 1982, earned designation on the National Registry. For more information call (254) 897-2611.

Story & photo courtesy Richard H. Moore.
Photo by Armstrongs' Western Fotocolor
Post Card Photographers, "Delux".

Dutch Windmill

Texas

As a tribute to the heritage of Nederland, Texas, settled by immigrants from Holland in 1898, citizens of the city erected an exact replica of a Dutch windmill. The design is one of many used in Holland. This forty foot tall structure houses a museum.

Artifacts include a trunk brought from Holland before the turn of the century, wooden shoes, the 1952 Gold Medal won by W.F. (Buddy) Davis, a native son, and mementos of the late Tex Ritter, country and western music star, who called Nederland home.

The first floor of the museum has a western theme, featuring a suit, boots and other items owned by Tex Ritter, presented by his family.

The Texas State Historical Survey Committee erected a historical marker in front of the building in 1970. A marker honoring War Veterans is located in a small park area owned by the city.

The windmill museum was a project of the Nederland Chamber of Commerce, started when an Englishman, M.L.C. Lucke, was president. Sam Bass, a German Engineer, built a scale model of the proposed structure and this was displayed in the Chamber of Commerce office. The model is in the museum.

After breaking ground for the structure on January 11, 1969, dedication ceremonies were held July, 1969, with the principle address by the Vice Consul of the Netherlands.

The windmill was built on a small tract of city property. Mrs. Dan Barras donated a city lot and the city purchased another lot adjacent to the original tract to form a small park area.

Artifacts are displayed on three floors of the building and there is a small gift shop.

For more information, call (409) 722-0279.

Story and photo courtesy of Cindy Clifton.

Shaw Gristmill

Texas

Although the Shaw Cabin, now located at the Log Cabin Village, Fort Worth, Texas, did not begin as a gristmill, both the structure and machinery found inside date back to the nineteenth century and are of historical importance. The Shaw Gristmill is also one of the few remaining working gristmills.

Well-known Thomas Shaw, a pioneer recognized for his log cabin building ability, built the cabin for his family in 1854. The original one-room house was considered one of the best-constructed log homes on the Texas frontier. Because of his exemplary skills as a carpenter and house builder displayed in his cabin, Shaw soon became widely known throughout the county. With his help many inexperienced homesteaders were able to erect log homes.

The 160-acre site on which Shaw constructed his house, preempted at fifty cents an acre, could be found on the extreme frontier in an area that came to be known as the Spring Creek community in Parker County, Texas. When first erected, not another pioneer settlement was west of this cabin, it sat only five miles away from a large group of Tonkawa Indians camped along the Brazos River.

After more than a century of use, first as a home, then as a bunkhouse, and finally as a storage barn, the cabin was relocated to the Log Cabin Village in Fort Worth. There it is being preserved and interpreted as a gristmill for educational purposes. A waterwheel and a stone wall were added to the original structure when the mill was set up. The overshot waterwheel is twenty-five feet in diameter, twenty-eight and one half inches wide.

The milling equipment, manufactured in the 1860s, came from a small mill, which had been in continuous use by the Smith family for over seventy years in the little town of Moline, Texas. During the middle to late 1800s, the family milled for farmers in and around Moline. After the turn of the century, the millstones were mostly idle, used only occasionally until 1930 to grind meal for the family or special friends. Then, forty years after it had completely stopped operating, the City of Fort Worth purchased and installed it in the Shaw home. Since 1970, the Shaw gristmill has delivered fresh, warm, and ready to use cornmeal, crushed by the same millstones and in the same fashion of over a century ago. For more information call (817) 926-5881.

Story and photo courtesy Ivette Ray.

Victoria Mill
Texas

In Victoria, Texas stands the sole surviving example of wind-powered milling in the region. The Victoria Mill is a wooden, semi-portable Dutch mill built by Fred Meiss, Sr., with the assistance of Otto Fiek, in 1870 near Coleto Creek.

The Dutch mill's sails are on a rotating turret with a long handle that reaches the ground, allowing the top of the mill to be turned so the sails always face into the wind.

Victoria Mill stands thirty-five feet tall, with four fifteen foot blades, a main shaft twenty feet long and fifteen inches in diameter. Wheels, sixty-five feet in circumference, drive the gears for the wooden shaft that moves the grinding stones.

Fred Meiss' great-granddaughter gave the gristmill to the City of Victoria in 1935. Restoration took place after the mill had been moved to Memorial Square (originally City Cemetery) in Victoria. Minor changes were made but the superstructure, shafts and log wheels are original. Although the mill is not used today, it does have the machinery necessary for grinding grain.

The Victoria Grist Windmill was placed in the National Register of Historic Places in Texas in 1979. For more information call Victoria Parks and Recreation Department.

Photo courtesy of Mrs. Gladys Arnold.

The Gardner Mill

Utah

Archibald Gardner was born in Kilsythe, Scotland in 1814. Imprisoned for speaking out against the Queen of England, Robert Gardner Sr., father of Archibald, packed up his family and sailed to Canada to homestead. Here Archibald built his first mill.

Archibald immigrated to the United States with his first wife, Margaret, and joined the first Mormon pioneers in their journey to the Great Salt Lake Valley in October of 1847.

After building several mills in the Cottonwood area of Salt Lake, Archibald began construction of the Gardner Mill, then known as the West Jordan Flour Mill, in 1853. The mill site also had a mattress factory, broom factory, saw mill, blacksmith shop, tannery, woolen mill and general store. It became the hub of economic activity in the southern part of the Salt Lake Valley.

The West Jordan Canal and many other irrigation canals were developed by Archibald and the first commercial water rights in Utah were issued for the Gardner Mill.

Gardner built over thirty-six flour, saw and gristmills in his lifetime. Among the last was the Star Valley Mill in Wyoming.

The Gardner family included eleven wives, forty-eight children and 270 grandchildren. Archibald eventually sold the West Jordan Flour Mill. John Aylett last used the mill for processing flour.

The mill and surrounding property passed to other owners and eventually became vacant. In 1979, Nancy Long bought the mill and converted it to a furniture store. Her dream of having a restaurant in the old mill came true in 1990.

The thirty-two foot cement silo was built in approximately 1905 and originally used for grain storage harvested on local dry farms. At full capacity, it held 1,200,000 pounds of wheat. That's as many as thirty-five truckloads each with 40,000 pounds of grain. That amount of grain would give the chef at Archibald's enough flour to bake bread for one hundred years. A sixteen foot overshot waterwheel was added in July of 1999.

In addition to the mill's renovation, Nancy and Chris Christenson, Joe Long (their son) and a hard-working staff, have found and moved historic buildings to the Gardner property to re-create a village complete with a stream winding through a series of connecting pathways and covered bridges. The Gardner Mill is listed on the National Register of Historic Places and received the 1987 Utah Heritage Award for the most improved commercial building. For more information call (801) 566-8903.

Story and photo courtesy of Marcia Banks.

Aldie Mill
Virginia

The Aldie Mill complex today consists of four buildings: the merchant mill, the granary, the country mill, and a house. Construction of the mill for noted legislator Charles Fenton Mercer began in 1807 under the supervision of William Cooke, a partner in the venture. The predominant brick structure, the merchant mill, and the house to the west were completed by 1809. The merchant mill ground wheat into superfine flour for commercial export. By 1816 Mercer's mill complex included a smaller brick country mill and a brick granary. Nearby farmers used the country mill to meet their needs.

The Mercer family demonstrated the feasibility of harnessing the Little River for a gristmill when they established a tub mill in the mid eighteenth century at or near the site of the present Aldie Mill. Charles Fenton Mercer decided to erect his mill at Aldie in order to capitalize on the Little River Turnpike, a thirty-four mile highway linking Aldie to Alexandria. When Mercer began construction in 1807, the Turnpike Company had completed their new road from Alexandria to the Fairfax County line. The mill's location also capitalized on the eastern junction of two other proposed turnpikes; one through Middleburg to Ashby's Gap, and the other to Snicker's Gap. With this network of roads, Mercer could count on gathering the lion's share of the plentiful flour trade from the rich farmland of the Loudoun valley. In addition to ample water power, good roads, and abundant supplies of grain, a state-authorized flour inspection station at Goose Creek conveniently served Aldie Mill. Mercer's flour, once inspected and packed, did not need to be handled again before leaving the country, thus substantially reducing costs and spoilage.

During Mercer's proprietorship Aldie Mill emerged as the largest manufacturing facility of its kind in Loudoun County. Originally powered by large wooden waterwheels, the mill incorporated the most advanced technology of the era. One of these early technologies was Oliver Evans' mechanical system, which moved grain and meal from one part of the mill to another without manual labor. Evans' machinery and the early wooden water wheels were replaced at the beginning of the twentieth century with roller mill technology and the present steel Fitz waterwheels and machinery. Much of the early machinery remains intact and on display, including the French flint buhrstones.

Among the mill's many prominent nineteenth century visitors were President James Monroe, who ground grain here while living at nearby Oak Hill, and Confederate Colonel John Singleton Mosby, who captured Union soldiers at the mill during the Civil War. Folklore has it that several Union soldiers hid in the flour bins and later, while fleeing, looked like ghosts.

Mercer sold the property in 1835 to the miller Captain John Moore, whose family operated the mill continuously for six generations until it closed in 1971. The Douglass family, descendants of Captain Moore, donated the mill to the Virginia Outdoors Foundation in 1981. Aldie Mill survives today as one of the best-preserved historic mills in the Commonwealth and Virginia's only known mill powered by twin overshot wheels.

For more information call the Aldie Mill, Virginia Outdoors Foundation @ (703) 327-9777.

Story and Photo courtesy of Andrea Brown.
Photo by Howard O. Allen.

BARGER'S MILL

VIRGINIA

Montgomery County, VA
Photo by J.C. Barger

Beverly-Chapman Mill

Virginia

Fire gutted the Beverly-Chapman Mill on October 22, 1998. The destruction of this six-story, stone, gristmill is an even greater tragedy because the mill survived a fire that gutted the building in the mid 1800s and another set by Confederate soldiers in the Civil War and an additional fire in July 1998.

The first three stories of the mill were built by Pearson Chapman in the 1740s. In 1858, the mill underwent renovations and the addition of the upper levels.

The mill, known as Chapman's Mill during the battle of Thoroughfare Gap, provided shelter alternately to Confederate and Union troops during the four-hour battle. Confederate troops occupied the mill at one time, using it as a meat-cutting warehouse and distribution center. Upon evacuation, the soldiers burned the mill so it could not be used by Union troops.

William and Robert Beverly rebuilt the mill in the early 1870s. Technological improvements were made in 1886. Steel rollers replaced the millstones, allowing the mill to keep pace with the production needs of the area, grinding nearly 100,000 bushels of grain annually. By 1946 the mill could no longer compete with mass production plants and had to close. The owners simply walked away, leaving the machinery in the building, all lost forever in the October 1998 fire.

Beverly-Chapman Mill is on the National Register of Historic Places. Only days before the fire, negotiations took place with hopes of restoring the mill to be used as a museum. Now, only the exterior stone walls of one of Northern Virginia's most historic places remains.

Old Mill News Winter 1999.
Watercolor by Joe Seme.

Big Otter Mill
Virginia

Lynchburg, Virginia Vicinity
Watercolor by Joe Seme

Brookside Mill
Virginia

Independence (Grayson County), VA
Photo by J.C. Barger

The Burwell-Morgan Mill

Virginia

The Burwell-Morgan Mill, a remarkable eighteenth century industrial landmark, was the fifth or sixth mill built by Nathaniel Burwell. Burwell, an entrepreneurial Tidewater planter partnered in this venture with a local resident, Daniel Morgan, a Revolutionary War hero.

Burwell envisioned a major commercial milling operation that would enable him to export the bounty of his lands, hard wheat, as high quality flour. He supplied the capital and Morgan the labor for the work begun in 1782. The mill was working by June 1785, making it one of the earliest merchant mills in the Valley.

Burwell sited the mill above Spout Run, a bold stream fed by five underground springs, which supplies enough flow to turn the great waterwheel (twenty feet in diameter) year round. The waterwheel is unusual because it is situated inside the mill - only six other mills in colonial America were built in that way.

Although mill technology evolved rapidly in the late eighteenth century, Burwell chose simple, proven techniques for the construction, taking advantage of the available material, straight-grained oak and close-grained maple. The sturdy hurst/husk frame the wooden gears and millstones are mounted on is set into bedrock to absorb the vibrations of the milling process and to keep the building from shaking apart. The worn hurst frame, waterwheel, wooden gears, and shafts are being replaced during the current renovation using the same materials and technologies that produced the originals. The great oak used for the main shaft came from Stratford Hall.

As the milling operation flourished, the village of Millwood became a significant export community. During the 1860 milling season the great wooden gears and immense millstones turned up to twenty-four hours a day, with five men working per shift. In that year more than 60,000 bushels of wheat were ground into flour. Flour from the mill was packed into barrels and transported overland by wagon or floated on barges down the Shenandoah and Potomac rivers to colonial shipping ports at Georgetown, Alexandria, and Falmouth. From there it was shipped to the Bahamas and then Europe. Allied businesses flourished. Local coopers, for example, built thousands upon thousands of barrels in which to ship the flour.

When the restoration is completed the mill will once again grind corn and wheat using the techniques of the eighteenth century. As one of the oldest agri-businesses in the Shenandoah Valley, this historic mill is still doing what it was built to do more than two hundred years ago. For more information, call (540) 955-2600.

Story and historical photo courtesy Clark County Historical Association. Color photo by James Holmes.

Bush Mill
Virginia

Bush Mill (aka Bond Mill)
Built 1907
Nickelsville (Scott County), VA
Photo by Stephen L. Spring

Cedar Forest Mill

Virginia

Cedar Forest Mill, built in the 1880s by Mr. Wills for John Hendricks, contained two runs of stone, one for wheat and the other for corn.

The mill sold in the 1900s to Mr. Moeschler, a blacksmith by trade. It changed hands again in 1914 when Captain Trent, a Confederate veteran, purchased the mill. Captain Trent installed Midget Marvel machinery.

The mill sold several times between 1914 and 1928. During that time it produced flour, cornmeal, sassafras oil and sumac for dyeing. The mill contained a blacksmith shop and grocery. It also served as a sawmill, making Cedar Forest a self sufficient village at the time.

In 1928 the Hubbard family built a concrete dam and employed Myrt Mann to run the mill for several years. Mr. Mann, who celebrated his 100th birthday in October 1998, still visits the mill from time to time.

In 1975 Edward "Ned" Lunsford purchased the mill from W.D. Sowell, after it closed for approximately twenty years.

Mr. Lunsford began repairs and restoration. He installed a twelve and one half foot waterwheel and sifter and replaced the chutes and elevators. He then cleaned and returned to running condition the machinery within the mill. Replacement parts came from various mills in the area no longer in operation.

In 1976 Cedar Forest Mill opened for business. At the time there were ten other flour-cornmeal mills in operation within a thirty mile radius. At present Cedar Forest Mill is the only one still producing flour and cornmeal in the area.

The mill contains three stands of rollers for flour and an original stone for cornmeal. The water that powers the mill comes from Straight Stone Creek. For more information call (804) 335-5136.

Story courtesy of Ned and Elaine Lunsford.
Photo by Stephen L. Spring.

3 Stand of Rollers

Cornmeal Stone

12½ Foot Waterwheel

Cleaning Machine

Cedar Forest Mill
Photos courtesy of Ned and Elaine Lunsford

Colvin Run Mill
Virginia

Colvin Run Mill Historic Site is located in Great Falls, Virginia. Listed on the National Register, it stands on what was once a major transportation artery connecting the rich farming valley of the Shenandoah to the busy maritime port at Alexandria.

In 1760 George Washington purchased a 275-acre tract of land along Difficult Run. By 1785 he had established a mill "seat" on the property, officially recording his intention to build a mill at the future site of Colvin Run Mill. Though it is not clear when Colvin Run Mill was completed, it is thought to have been built about 1811 by William Shepard on land purchased from Washington's heirs. Philip Carper bought the mill from Shepard in 1811 and is the first documented owner of a gristmill business here. He and the following owner, John Powell, hired millers to operate their merchant mill enterprises.

Colvin Run Mill was closely modeled on the technology developed by Oliver Evans (1755-1819), whose automated process revolutionized milling in late eighteenth century America. In Evan's system, mills became modern factories where gravity and waterpower were used to move grain and meal through the milling process.

In 1883 Addison Millard, a miller from Frederick County, Maryland, bought Colvin Run Mill. The Millard family owned and operated the mill until it ceased operations in 1934. After Addison's death in 1897, his widow, Emma, carried on the business, and left it to sons, Sam and Alfred, upon her death in 1919. During this time the millpond and millrace were rebuilt and the mill machinery modernized. The greatest change was in the grinding process. A new machine, called a roller mill, proved more efficient (but not necessarily better in quality) than the "old fashioned" millstones.

In 1965 the Fairfax County Park Authority acquired Colvin Run Mill and began a full-scale restoration to replicate the original Oliver Evans design. Standing as an example of early America's technological innovations, Colvin Run Mill is an award-winning masterpiece of restoration craftsmanship. The work of many skilled hands have literally recreated the past.

As in earlier days, the mill is run by waterpower. Water from Difficult Run, diverted into a millrace, flows over a twenty foot overshot waterwheel hand-crafted of white oak. Today, Colvin Run's miller once again uses French buhrstones to grind flour and cornmeal. Colvin Run General Store (ca. 1900), the early nineteenth century miller's house and an interpretive barn with adjacent blacksmith shop are also located on Colvin Run Mill Historic Site. For more information call (703) 759-2771.

Story and photo courtesy Ann Korzeniewski.
Photo by Don Sweeney, FCPA.

Cowan's Mill
Virginia

Stone Mountain (Lee County), VA
Photo by J.C. Barger

Forestville Mill

Virginia

Forestville Mill, formerly known as the Andrew Zirkle Mill, has stood on the banks of Holman' s Creek in Shenandoah County, Virginia, since approximately 1757. The village of Forestville owes its existence to previous mill owners who subdivided the land and sold building lots.

The mill operated throughout the Revolutionary and Civil Wars. Local legend tells of occasional Indian attacks. Early residents of Forestville would have taken refuge in the fortress-like mill. Apparently the builders planned for such occasions by leaving narrow slits in the stonework wide enough for muskets.

During the Civil War, the mill was one of the few Shenandoah Valley structures to escape the torch of Major General Philip Sheridan's Union Army. In the fall of 1864, Sheridan's soldiers set fire to another mill; Samuel Hockman, miller of Forestville Mill, hoisted a Union flag and saved the mill but met with disfavor among the local citizens!

In *Life Along Holmans Creek*, by Joseph Floyd Wine, the writer states that despite their feelings, "The destitute families sought credit from Mr. Hockman stating that they had no money to buy flour. Unreceptive to their pleas and responding to their question, "How are we going to feed our families?" he replied, "You will have to bake cakes." After the war when other mills were rebuilt, the miller found himself without financial income, as many local families would not deal with him. In his unfavorable circumstances, he sought understanding from the residents saying, "what am I going to do?" One unsympathetic farmer answered, "You will have to bake cakes."

In the early 1930s, during the great Depression, another Zirkle, Paul J. Zirkle, purchased the mill and tried to make it prosper. One venture was a cider press rigged to use the Fitz waterwheel for power. Orchard owners gave away apple drops, and local citizens would collect them and deliver them to the press by the wagonload. Mr. Zirkle would accept a portion of the cider in exchange for the pressing. Another endeavor was a 100-mile flour route across the mountains into West Virginia, then north before turning back into the Valley.

In spite of these efforts, modernization ended the Forestville Mill usefulness, and the mill closed in the 1940s. Robert Eaton owned the mill after its closure and for a time kept chickens in it, screening off the old mill machinery. The mill sat derelict until 1980 when Glen Hofecker purchased it and began its first renovation. In the 1980s, Mr. Hofecker built long case clocks and Chippendale furniture at the mill.

Gordon D. "Sonny" Bowman, II purchased the mill at auction in October 1996. While he had no immediate plans for the structure, he recognized the need to save the mill for future generations. Externally, the mill is a very impressive structure with limestone block foundation, weathered cedar siding and metal roof, but stabilization and restoration has been required.

Side view with flume and water wheel

Inspection of the Forestville Mill by millwright Derek Ogden in June 1997, identified and explained the use of basement wheat sinks, Howes grain cleaners, line shafts, elevators, Hurst frames, and the twenty-one foot Fitz waterwheel. There are two pairs of French buhr millstones in excellent condition. A stone crane and roll stand, manufactured in the 1890s, are on the first floor. The second floor contains wheat bins, corn bins, a Barnard and Leas plansifter and a homemade dust collecting room. The third floor, or attic, holds dressing machine reels, a cleaning reel, and elevator heads with sprockets and link chain drives. At the end of the day, Ogden inspected the dam on Holman' s Creek. Recent floods have damaged the headrace and the banks have been scoured. Most important, however, there is an assured water source controlled by Mr. Bowman. In so many cases throughout Virginia, the structure may still stand and milling machinery exist, but the water source is gone. Many restored mills must rely on municipally supplied water or pumps and long runs of pipe. As late as 1985, the Forestville Mill's waterwheel operated using water supplied by dam, millrace, and wooden flume!

"Footprints" of previous mill technology are found on all three floors and in the basement. Mills built prior to the 1790s had no (grain) elevators and the basement would be the main working area for the miller. Most adjustments were in the basement until technology improved and millers moved the work area and millstones to the first floor. The mill machinery has probably been replaced (upgraded) two or three times, with the latest equipment installed in the 1890s and 1920s. The "footprints" Mr. Ogden looked for are holes cut (and later patched) in floors and ceiling beams, ceiling joists cut, and holes in the massive stone foundation. While there is no stone chimney serving the large basement fireplace, Mr. Ogden believes it may have been replaced with a smaller brick chimney at the same time changes were made in the attic and roof. Fire or storm damage, or the installation of new machinery, could have necessitated this. Mill floor space was probably at a premium and the large stone chimney that sat in a corner was later turned into an office.

With the detective work and repairs accomplished, it remains to be seen to what use the mill will be put. In the meantime, an important part of the history of The Shenandoah Valley is kept for future generations. Mr. Hofecker, and later Mr. Bowman, have done a wonderful job of restoration.

Story by Kinter McClelland.
Photo by Joe Seme.

Forestville Mill
Photo by Joe Seme.

Gate City Mill
Virginia

(aka Beverly Mill)
Gate City (Scott County), VA

Hollen Mill/Dry River Mill

Virginia

Dry River Mill is typical of late nineteenth century gristmills in Rockingham County. Constructed in 1874, the mill building is a three-story rectangular structure with additional space in the garret. The main entry in the three-bay, east gable end, had a central door flanked by windows. Although the north fenestration has been changed on the first floor to accommodate a modern apartment, the upper levels display two bays with 6/6 sash. The pen stalk sheltering the turbine projected off the west gable end; this consisted of a two-story frame shed, which did not continue the length of the west wall. Like many of the local gristmills, Dry River Mills also ran a sawmill, which stretched along the south side of the taller gristmill. This narrow one-story structure, later appended by a frame addition on the west , is believed to have been brought to this site from a cider mill further up the river.

The interior of the mill still displays its large hewn timbers and chamfered posts, mainly of yellow pine, on each floor. Horizontal board siding finished the interior walls on each floor. The first floor has been remodeled into an apartment, retaining some of its original finish. The hominy mill and corn crusher were originally located on this floor, as are the timbers that supported the millstones.

The second floor contained the office in the southeast corner, opened up when the mill was converted into a woodworking shop. The flour chests, millstones, and room where the flour was barreled were all located on west end of this floor. Scales to weigh the wheat had been located over the front door.

The third floor primarily provided storage area with bins along the south wall and the northeast corner. The bolting machine, which sifted the flour, occupied this floor near the center of the room. The present owners added the Dutch door on this floor.

A group of local men, Abraham Miller, J.W.F. Allemong, A. R. Hollen, John L. Kerlin, and Eli Hollen, petitioned for a charter for a flour and a sawmill March 30, 1871, to be located at the "Miller Factory Property" on Dry River. They applied to the state again in 1874 with Daniel Miller added to the list of petitioners and received their charter of incorporation as the Bridgewater Milling Company. Abe Miller deeded the land, nineteen acres, to the corporation September 1874 (DBll:185). Information was taken from the Historic Landmarks Commission survey form.

The mill was apparently built in the fall of that year. By December 26, 1874, Abe Miller had rented the "new" mill to J.T. Wright and N.Y. Partlow, currently doing the milling business known as J.T. Wright for two and one half years, beginning January 1, 1875. The millers promised to "make a 1st class fancy brand of family flour," to put in a "middling purifier," and to pay said Abe Miller $700 a year. Miller agreed to complete the miller's house "now under construction."

According to Carrie Hollen Simmons, the Bridgewater Milling Co. later sold the mill to Nancy Miller Hollen, daughter of Abe Miller. She left the mill to her husband, Adam R. Hollen. After his death around 1912-1913, his three sons, Abe, Luther, and Tom, acquired the mill. Tom actually ran the mill, and his two brothers sold their interests to him. He ran the mill until about 1927. After Harrisonburg began to get water from Dry River, Hollen didn't have enough water to run the flour mill, and he could only run the sawmill when the water level was high. Some of the older mill workings are currently housed on the upper floor, including many gears, shafts, wooden augers, and grain elevators. The upright gear and shafts, which ran the bolting chest, are also clearly visible. The long wooden box extending along the ridge contains the wooden auger, which ran the hoist to carry flour up to drop it in the mill. The sawmill is currently used for storage.

The miller's house "under construction" in 1874, has been extensively remodeled in the 20th century. Ray and Carrie Hollen Simmons bought the mill after her father's death in 1954. Leonard and Elsie Hollen pictured in front of the mill are Carrie's brother and sister.

Story and photo courtesy Janet Downs.

Jesse Mill
Virginia

Dillard (Russell County), VA
Photo by J.C. Barger

Mabry Mill

Virginia

E.B. Mabry built and operated Mabry Mill from 1905-1935. First used as a blacksmith shop, Mabry later added a sawmill, gristmill and woodworking shop. Today, as in the past, water still powers the wheel to produce cornmeal.

Mabry Mill is one of the most photographed mills on the Blue Ridge Parkway. The appeal is so great, pictures of Mabry Mill are seen on postcards representing rural life of states other than Virginia.

The Mountain Industry Trail, on which Mabry Mill is located, is an easy, educational walk of less than a half of a mile. There are buildings, farm implements, a whiskey still, and other displays that illustrate what rural life was like in the Appalachia and other areas of the eastern United States. At times there are demonstrations of mountain crafts such as gristmilling, spinning, and blacksmithing, which were a part of the early 1900s. Also, there are recipes for homemade soap and moonshine.

On the pond turtles, ducks and other wildlife can often be seen. All of this makes the Mabry Mill Trail one of the more popular trails on the Blue Ridge Parkway.

Story courtesy of Joe W. Getsinger.
Photo by J.M. Tanko, Blacksburg, VA.

Matthews Tide Mill

Virginia

Tidewater, Virginia, is home to an eighteenth century tide mill. The gristmill ground meal and flour for the Continental Army. The great wheel is turned by the tide of the East River in Poplar Grove, Matthews County.

Story by E.P. Johnson.

Photo courtesy Society for the Preservation of Old Mills.

McCormick's Mill
Virginia

Steele's Tavern (Augusta County), VA
Photo by Steven L. Spring

Meadow Run Mill

Virginia

Approximately two hundred years ago the Michies operated a mill that unfortunately fell from decay. In order to keep tradition alive and recreate the effect of eighteenth century life, the owners of Michie Tavern (ca. 1784), moved and reconstructed the Meadow Run Mill, once located at Laurel Hill, Virginia. The appearance resembles the mill originally established by the Michies. The General Store housed in the Meadow Run Mill depicts the mercantile atmosphere of Longwood, the store the Michies ran.

The history of Meadow Run Mill remains vague. The first claims of its existence come from the ancestry of the Coiners, who were homesteaders in Crimora during the 1750s. The ancestors spoke of two mills in operation at that time, one on Christians Creek destroyed by fire several years ago and the other on Meadow Run.

It is believed the plans for the mill were drawn up in 1764 and construction began in 1770. The site for the mill was cleared and the mortar for the foundation prepared. Never completed, legend has it that the site served as a military depot where armaments were stored during the Revolutionary period.

Another legend tells of the mill beginning as a coffin factory, the construction having been concluded prior to 1797. The owner and operator at that time remains somewhat of a mystery. Supposedly, the owner of this coffin factory would pluck a "horse reed" from the banks of Meadow Run and travel to the home of the newly deceased to measure the length of the body. He would take the reed back to the factory and use it to determine the size of the coffin. Once the coffin-maker used the reed it would placed in front of a window that faced the road. Local townspeople on their way to the marketplace would stop to inquire of the deceased. Upon arrival at the marketplace, the villagers would then spread the news.

One story tells of possible tragedy and certainly creates an atmosphere of mystery. During the early 1800s the owner of the mill had a lovely fair-haired daughter, who was very much in love with a young apprentice-miller. Her father did not approve and the lovers began to meet secretly, primarily at night. The girl would stand on the balcony over the wheel of the mill and flash a coin to signal when her father went to sleep. The apprentice would flash a similar coin in acknowledgement. He would then cross the stream to the wheel. The young man's sweetheart would lock the turning wheel by wedging a piece of timber between the hub and the mill wall and he would climb up the wheel onto the balcony. One night the wedge of wood slipped, unlocking the massive wheel. Because of the man's weight, the wheel began to turn very quickly and the maiden tried to save her lover by leaning over the balcony and grasping his hand. She was torn from her stance over the railing of the balcony. The following day coins were seen glistening from the stream in the early morning sun. It is believed by some that the lovers perished. Others claim the young couple ran away together. However, to this day, it is said that during the light of the moon one may still see the flashing of coins and hear the calls of the lovers.

As well as a site for grinding operation, the mill served as a gathering place for area residents. The women would meet with their sewing or produce and make fine quilts stuffed with goose down, or jams and preserves. Local fiddlers, harmonizing with banjo and guitar players would provide music for the festivities. An elderly gent once spoke of whiskey being made in a distillery set up at the mill. Though not proven, he emphatically said entertainment of friends required some kind of drink and with corn and rye so readily available, "moonshine" was bound to be one of the miller's products.

Between 1821 and 1855, the mill sold several times. In 1856, Windle and Mary Sites purchased this property and continued running the milling operations. It appears the Sites, who owned the mill during the Civil War, were most pleasant and kind in nature. Windle and his wife, great sympathizers of the slaves quest for freedom, were pacifists by way of their religious beliefs. When Yankee troops came to burn the mill down, Sites convinced the commanding officer that the slaves were free and received monthly wages. The leader of the troops found the claims to be true and decided to spare the mill. The officer later made arrangements for the mill to be used as an interim hospital for Yankee soldiers.

Later, during the autumn of 1888, Windle Sites sold the mill. After changing hands a few more times, John Drumheller purchased the mill in 1896. His son, Carson Drumheller, was the last person to run the mill, keeping the operations intact until 1958. In 1974, the Meadow Run Mill was procured by owners of Michie Tavern (ca.1784) for the purpose of restoring and preserving this unusual aspect of American heritage. For more information call (804) 977-1234.

Story courtesy of Cynthia M. Conte.

Meadow Run Mill
Photo by Stephen L. Spring

The Mill at Carter Hall

Virginia

Millwood, Virginia is unique on the American mill scene. It is home to two restored mills built by the same man in the late eighteenth century - one was a merchant mill, the other a plantation mill.

Six hundred yards down stream from the Burwell Morgan Mill on Spout Run in Millwood is "The Mill at Carter Hall". Built as a plantation mill in the late 1790s by Nathanial Burwell, owner of the 5000 acre Carter Hall Plantation, it was originally known as Spout Run #2 or the Lower Mill. It later came to be known as the Boysteyan Mill for Adam Boysteyan, the original miller, and has recently been renamed the Mill at Carter Hall by new owners Roger Chavez and Susan Fielding.

The mill was built and operated by Burwell's slaves to grind wheat and corn for the Carter Hall Plantation. The stone was quarried on the site and cut as needed. The 450-yard headrace became a monumental construction project. Some of the race cut through solid rock and the earthen dikes remind one of a canal rather than a millrace. The race incorporated a dam, gates, culverts, flumes and spillways. The new owners are presently clearing eighty-five years of reforestation and new growth from the race but there is no intent to reactivate the race. It will become a historic nature walk and a testament to the energy and tenacity of the slaves who built it. Many of the descendants of these men still live in the Millwood community.

Sometime after Nathanial Burwell's death in 1814 the mill sold and became a merchant mill grinding for the local community and export to the West Indies and Europe. It passed through a succession of owners and did not operate during World War I. The building was restored in 1987 and will soon become the home of an antique letterpress print shop and art gallery devoted to the letter arts.

Story and color photo courtesy of Roger Chavez.
Historic photo courtesy
The Handley Library, Winchester, VA.

MOCK MILL
VIRGINIA

Meadow View (Washington County), VA
Photo by J.C. Barger

Mt. Airy Roller Mill

Virginia

In 1830 William C. Miller built Mt. Airy Roller Mill on Stinking River in Pittsylvania County, Virginia. W. Posye Johnson purchased the mill, then known as Cork Mill, and twenty-one and a half acres of land in 1917. This mill has large wooden pegs and Roman numerals on the ends of timbers, indicating that the timbers were sawed elsewhere and brought to the site to be assembled. The building is three stories high and has two levels in the basement.

William Miller ground wheat and corn on two stones. He also had a cotton gin, sawmill, wheelwright and blacksmith shop and foundry to make plow points, wagon wheels and other metal parts in this area.

A Crown grant of 2400 acres of land to the Crenshaw family provided a place for the grain to be grown. William C. Miller purchased the property in 1820 and remained owner for eighteen years before deeding it to his sons, Nathaniel and Charles E. Miller, in 1838. They continued the operation of the farm and mill until 1914 when the property sold in tracts.

The mill had various owners until 1917 when purchased by W. Posye Johnson, a third generation miller. The mill was reworked and Salem flour rolls and a Fitz overshot waterwheel were installed in 1920, leaving one stone and water turbine to grind cornmeal and feed until 1945. A rock and wooden dam had been built on the property at the time of purchase. After a flood in 1928, half of the dam was replaced with concrete and the other half was replaced in 1945 after another flood. Other damaging floods came in 1987, September 1995 and Hurricane Fran in September 1996.

After the death of W. Posye in 1948, his son, William P. "Billy" Johnson accepted operation of the mill and ground flour from 1948 to 1968, but poor farm economics made it necessary to discontinue grinding flour. In 1952 Billy added a Hammermill to custom grind and mix livestock feed powered by diesel engine. Billy has operated the mill for fifty-one years and currently operates the only feed mill in Pittsylvania County.

Story and photo courtesy Billy and Emma Jean Johnson.

Osceola Mill

Virginia

Steele's Tavern - Marl Creek (the name derived from loose lime-rich deposits that form portions of the stream bed) is normally a narrow, clear stream with a strong and steady, year-round flow of water fed by numerous springs along its course.

It arises near the old McCormick Homestead, site of Cyrus McCormick's invention of the mechanical reaper, and is now a wayside picnic area owned by the Commonwealth. It flows for a little more than a mile by adjacent roads into South River, hence to the Maury and James rivers and on into the Atlantic Ocean.

The strong and reliable flow of the creek provided for many decades a welcome source of power for local industries before the development of steam and electricity as superior sources of energy.

There were at least four and perhaps as many as seven waterpowered mills along this short strip of water at various times extending from the late 1700s into this century. It is said that Marl Creek supported the largest concentration of mills in the shortest span of water in Virginia for many years.

Three or more of these mills were built and owned by the McCormick family, beginning with the original homeplace mill near the headwaters. Another was owned by the Steele family for generations. The Steeles were early area landowners and founders of Steeles Tavern.

Two of these mills stand today, the homeplace mill and the last one on Marl Creek before it empties into South River about 300 yards below - the Osceola Mill

Osceola Mill was built by Cyrus McCormick in 1843. It is laid on a solid limestone foundation and its heavy beamwork and much of the siding is of original native chestnut.

Water propelled the twenty-eight foot Fitz overshot wheel via a large pipe from a millpond some distance above the mill. The stone-dammed millpond, easily seen from the road may be the only one remaining along the creek.

It is thought that the original millwheel was smaller and fashioned of wood and that the current wheel dates late nineteenth century. While the mill is not operable today, the mechanical components are still on the premises and could be restored to working order.

Today, as a country inn, the Osceola Mill is a wonderfully rustic relic with a relaxed and intimate atmosphere that belies its size. For more information call (540) 377-6455.

Story and photo courtesy Brian Domino, Inn Keeper.

Stratford Mill

Virginia

The Stratford Mill sits on the bank of the Potomac River in Westmoreland County, Virginia on Stratford Hall Plantation. The ancestral home of the Lees of Virginia and birthplace of Robert E. Lee, Stratford Hall was built in the late 1730s by Thomas Lee (1690-1750). In 1743, Lee added a critical tract of land near his landing on the Potomac River. The deed of sale, recorded on 14 March 1743, states in part "Waddey, Jeane of Northumberland County, Virginia, widow of Benjamin Waddey to Thomas Lee, one acre situate in the County of Westmoreland and at both ends of an old mill dam near the said Lee's landing and bounding on the dwelling line between the said parties that is half an acre at each end to be marked and bounded by the patents hereafter in the most convenient manner for a mill."

Although the exact date is not known, it is believed that soon after acquiring the property Thomas Lee constructed a mill (Lee Mill) on the site. As the deed of sale states, "an old mill dam" existed at the location. No details concerning the earlier mill are known but a county survey of 1721 lists a mill at the "Cliffs." Cliffs Plantation was built by the Pope Family, on the land that later became Stratford Hall Plantation.

The Lee Mill with its advantageous location proved to be productive and profitable. In 1759, an act of the House of Burgesses designated Stratford Landing as the official shipment, storage and inspection station for tobacco in the region. This designation increased commercial activity along the waterfront. Although no records from the mill have survived, the increase in shipping traffic at the landing would have afforded the Lees the opportunity of stocking ships with flour and meal for the long journey back to England.

The years of commercial success for the landing and mill suffered a setback in 1769. A hurricane destroyed the tobacco warehouse and wharf. Blocked by political foes, the Lees were unable to renew the license for the tobacco warehouse. Repairs were made to keep the mill operational. Inventories of Philip Ludwell Lee (eldest son of Thomas Lee) list mill repairs in 1775, 1776 and 1778. The inventory of 1780 notes the purchase of a millstone and in 1781 payment to a millwright for work at the mill. By 1860, the mill was no longer operational and in poor condition.

In the 1860s, Messrs. Muse and Jenkins rebuilt and leased the mill from Elisabeth McCarthy Storke, then owner. The mill remained operational until 1906. During the next thirty years the mill laid silent and suffered under the elements.

After years of private ownership, Stratford Hall sold to the Robert E. Lee Memorial Association, Inc. in 1929. The Association decided to restore the house and mill. Mill reconstruction began in 1939. All that remained of the old Lee Mill were parts of three stone walls and the foundation. Under direction of architect, Fiske Kimball and Resident Superintendent of Stratford Hall, General B.F. Cheatham, many skilled individuals were brought into the project.

James F. Bell, of General Mills Corporation, commissioned Professor B.W. Dedrick to conduct research on eighteenth century mills then acquired and installed mill machinery located by millwright, Alden Ackles, in Maryland and Virginia mills. Senator and Mrs. Metcalf of Rhode Island donated funds for reconstruction of the mill and dam. The Fitz Waterwheel Company rebuilt the rolling screen and bolter and constructed other major components of the mill including the hurst frame and waterwheel. The reconstructed mill, dedicated October 1939, remained in operation unless repairs were being done.

The Stratford Mill is still operational and there is a Plantation Store on the site. For more information call (804) 493-8038.

Story and photo courtesy of Steven T. Bashore.

TURNER'S MILL
VIRGINIA

Broadway, Virginia Vicinity
Built early 1800s and survived the Civil War.
Watercolor by Joe Seme

GEORGE WASHINGTON'S GRISTMILL

VIRGINIA

George Washington's Mount Vernon plantation was an ambitious commercial enterprise. In the 1760s Washington replaced tobacco with wheat as his staple crop. He made the plantation more self sufficient by manufacturing cloth and expanding his fishery to supply the food needs of his slaves and to provide another marketable commodity. He implemented a variety of measures to improve the fertility of his soils. In 1771 George Washington replaced a severely deteriorated gristmill erected by his father, Augustine, decades before. The new mill, outfitted with two pairs of millstones including a pair of French buhrstones, would grind the grain of neighboring farmers and that grown in Mount Vernon. According to an 1803 insurance document, the mill was built of stone, two and one-half stories high, thirty-two feet by forty-six feet in dimension. In 1791, Washington had a series of improvements developed and patented by Oliver Evans of Delaware installed to the millworks. Washington appears to have been one of the first to adopt Evans' labor saving system.

Washington often experienced difficulties in engaging a competent miller. The mill also suffered from a lack of water in the summer months and a long string of accidents, including the collapse of the wheelpit walls. The mill still proved to be a highly profitable venture. Washington marketed his flour widely, shipping it as far as the West Indies. In 1797 Washington's farm manager, a Scot named James Anderson, convinced his employer that producing corn and rye whiskey would be a natural complement to the milling operation. Anderson supervised construction of a substantial distillery, a stone building seventy-five feet by thirty feet in dimension, with five stills. In 1799, the second year of the distillery operation, almost 11,000 gallons of whiskey were produced and sold for $7,674. The mill and distillery fell into disrepair in the nineteenth century and both were completely dismantled by ca. 1850. The Commonwealth of Virginia reconstructed the mill in 1933 and has operated it as a state park since that time. The reconstruction has been based on documentary evidence and the results of archaeological investigations that revealed the footprint of the original mill. Evidence indicated that the wheel pit was located inside the building. A fragment of the waterwheel was also found, preserved in place in the moist, oxygen-free environment at the bottom of the pit. The reconstruction included reusing gearing and other elements removed from an early nineteenth century mill located near Front Royal, Virginia. At present, the reconstructed mill is undergoing restoration, with the goal to return it to full operation by Spring 2000. Archaeological investigations are underway to locate the cooperage, slave quarters, and animal enclosures that were located at the site. Portions of the foundations for the distillery have been revealed and archaeological excavations to learn more about that important element of the mill complex are ongoing.

Story and photo courtesy of D.J. Pogue

White's Mill
Virginia

Deeds and historical records indicate the existence of a mill built on this property by Thomas Moffett in 1797. The millstones were laid in that year by Joseph Lewark. It is quite possible that the stones in the foundation of the present mill are from that earlier structure, but there is no positive evidence that the site is the exact location of today's mill.

In 1866, based on a lease agreement signed in that year between William Y. C. White and James Ireson, Ireson was to build a "mill house" during the five years of the lease. However, the document goes on to say that "After completion of the mill house and repairs made upon the mill next season..."

In 1838 Colonel James White acquired this tract of land, on which existed a "water gristmill and carding machine." Tax records for this period indicate a very low, depreciated value. By 1842 there was an increase of $1,150 in appraised value, which remained consistent through 1875.

It would appear from information available, that the current mill probably dates around 1840, with parts of it dating to an earlier period. Because Col. James White died in 1838, shortly after purchasing the property, his son, William Y. C. White, who received the mill property in the 1856 partition of the estate, probably built the mill.

White's Mill is an integral part of the historical character of "Southwest" Virginia and the Abingdon Historic Area. A full-time millwright maintains the historic nature of the mill and demonstrates its operation.

The old general store and the "Mill House" (home) originally a part of the mill property are adjacent to the mill. Additionally, the entire area adjoining the Mill encompasses numerous historic structures predating 1900. Both the Mill House and Old Store are being evaluated for consideration of distinction as historic structures.

Story courtesy Eric L. Morgan.
Photo by J.C. Barger.

Barron Flour Mill

Washington

The sign at the Oakesdale Mill says, "If no one here, serve yourself." For decades, customers left their money and took their flour and a bit of Joseph Barron's soul in every bag.

For more than seventy years, the miller of Oakesdale ground wheat into flour, feeding first the farmers of the Palouse, then their livestock and finally organic diners.

Joseph's father, Moses Barron of Barronsville, Pennsylvania, where Joseph was born in 1866, owned and operated a water-powered gristmill. Joseph wanted to follow a different line than that of his father, and studied for the ministry. But after finishing theological school, he decided after all to go into the milling business and worked with his father in Pennsylvania.

In a few years Joseph married and took over a mill of his own in Crawfordsville, Indiana, operating it for three years. The westward movement was well under way by then and Joseph and his wife Martha joined a group coming to Aston, Washington.

Soon after, in 1907, he purchased the mill in Oakesdale. It was converted to electricity. Waste lumber used to stoke the furnace of steam boilers and the addition housing the steam generator became living quarters for his family. The mill had a capacity of eighty barrels a day. A barrel of flour equaled 196 pounds, and a sack of flour weighed forty-nine pounds.

Around 1916, Mr. Barron "traded his mill" for large farm acreage in Montana. Quoting his son, it "turned out to be a fiasco." In 1919, Joseph Barron again opened his flour mill and started over. The mill made flour steadily until 1939. Flour milling became unprofitable, but the Barrons held on by cleaning seed, storing grain and selling coal.

In the 1940s the elder Joseph turned his mill over to the management of his son, young Joseph. The old mill operated until 1960. Calendars from those years still decorate the mill. Notebooks with farmer's names scrawled in pencil hang on nails above bins.

The old mill stands along McCoy Creek, where the willows dip and bend. Built in 1890, it is three and a half stories tall, full of intricate well-kept milling equipment, everything clean and neat. For years, Barron's products sold under the trade name "Nutri-Grain" until Kelloggs bought the name in 1980.

In a new mill across the creek, Barron milled 500 pounds of flour an hour. But at eighty-eight, hefting fifty pound sacks of flour became too much. Maintaining the immense old mill, not operating since 1960, seemed impossible. He considered giving it to one of the area's historical societies, but they were too distant or too poor to pay the upkeep. So Barron put his mills up for sale, sifting through prospective buyers as carefully as he has different kinds of wheat. According to his only child, Joan Roehl of Kenmore, Washington, dozens of people inquired.

December 4, 1998, the only surviving flour mill of the 1800's once gracing Whitman County sold. Longtime customer Mary Jane Butters and her husband, Nick Ogle, agreed to purchase the old mill. Butters, an organic farmer from Moscow, Idaho, needed a place to grind garbanzo beans. Barron agreed to let her use his equipment. Both his new operation and the old place immediately enchanted her. She loved the old mill's woodwork, handrails worn smooth, the knot-free lumber and the function of the mill as glue that once brought the community together. "The first time I toured the old mill, I was humbled by the grace and grit of their lives."

Barron had just used a grant and $10,000 of his money to put a new wooden shingle roof on the mill, which is listed on the National Register of Historic Places. He spent about $1,000 a year on upkeep of the old place. But it wasn't until his health started to go that he wanted out. "I hate to do it. I've made many friends and had a lot of good customers who expected good flour. They continued milling Joseph Barron's flour using his new (1960) mill equipment. They kept the name, "Barron Flour Mill," also the three-grained hot cereal he invented and even the hand-lettered 'self-serve' sign. To Butters, it's a perfect extension of her mail order company and her goal of producing healthy food. To Barron, it was simpler. "She liked the old stuff and was willing to keep it," he said.

Barron's wife, Ethel, died more than a decade ago. He lives near the two mills with his cat, Blackie. Barron wanders out some mornings to the new mill, where, as Ogle putters, flour dusts Barron's shoes and cheeks. "Never lost the place," Barron says. "Never lost it."

'Old Mill Days,' Oakesdale's annual community celebration, is based around the mill. "It is there in the middle of town reminding people they once were a community relying on each other, Butters said. For further information call (208) 882-6590.

Story and photo courtesy Mary Jane Butters.

Cedar Creek Gristmill

Washington

The Cedar Creek Gristmill sits on a rocky ledge above the gurgling waters of the creek that gives the mill its name. The gristmill rests at the bottom of a narrow tree-lined gorge, near the old community of Etna.

On July 22, 1876, the Vancouver Washington newspaper, *The Independent*, reported, "George W. Woodham is building a gristmill on Cedar Creek. It will be ready to grind wheat this fall." It is rumored that the mill had a wooden waterwheel. Water turbines, such as the sixteen horsepower James Leffel presently powering this mill date 1862, lending credence to the possibility of turbine power from the beginning.

The mill evolved into a center of activity for the people of Etna and the surrounding countryside. They would gather for music, dancing, and swapping the latest news or gossip. They would often camp on the other side of the creek from the mill or just socialize with friends they had not seen for some time.

Woodham operated the mill as the "Red Bird Mill" for three years. The mill sat idle for seven years. Mike Lynch then purchased the mill and leased it to Gustave Utter in 1888. After reinstalling flour milling and feed machinery, Utter operated the mill until 1891. He also moved on because he could not run the mill profitably. The mill may have been run as an informal co-op from 1892-1905. Goran Roslund purchased it in 1905.

By 1909 Goran had added a shed to the upstream side of the mill used as a cedar shake mill and a shed on the downstream side used as a blacksmith shop by his son, Elmer. Victor, another son of Goran, a mechanic, turned the main floor into a machine shop.

Victor Roslund, an individual of many talents, worked on an aircraft engine and generated electricity for his and his relatives use years before it was available to other residents. Parts made or repaired at the mill are still in use at Merwin dam. Victor is primarily responsible for the survival of the mill. Without his desire to keep it running it would have surely fallen into disrepair many years ago. Victor lived on site in a room upstairs with his mother until the early 1950s. He sold the mill to the Washington State Department of Fisheries on January 1, 1955.

Prior to the sale the cedar log dam was removed from the creek as the first stage in the construction of a fish ladder. It is believed that it took three tries with dynamite to remove the dam. During the late 1950s and early 1960s the mill fell into disrepair.

In the mid to late 1960s, the Vancouver Historical Society leased the mill and did a partial restoration. The work done by that organization kept the mill from sliding into the creek. Jeff Bizzel also worked to save the mill in the 1970s but the mill suffered from neglect, rot and vandalism. By 1980, the cedar shake mill shed had collapsed from heavy snow. The mill's main flooring was unsafe and the east wall had all but rotted away.

Friends of the Cedar Creek Gristmill formed in 1980. The group of local residents dedicated themselves to saving and restoring the oldest surviving original commercial building in the state still operating. No drawings or photos of the original layout were ever found, so guidelines typical of a 1900 era gristmill were used.

Work parties replaced the old metal roof with a new cedar shingled roof. Rotted beams and boards were removed and replaced with new hand-hewn ones prepared with tools of the past, such as broadaxes, adzes, and peeling spuds. Concrete and cables strengthened the foundation and lumber resembling the original replaced the old floor. Wooden windows, made by hand, were installed. Beams were put together with mortise and tendon joints held by wooden pegs.

Rocks were blasted from the creek bank and a 650-foot flume of treated lumber and steel or concrete was constructed. Stones were embedded in exposed concrete to blend with the surrounding environment.

The goal of grinding grain on Washington State's Centennial Birthday took place on November 11, 1989, at 3:00 pm. Demonstrations of the milling process and restoration continue. For more information call (360) 225-5832.

Story and photo courtesy of Fred Schulz.

Bunker Hill Flour Mill

West Virginia

Along the banks of Mill Creek in Southern Berkeley County, West Virginia, stands the Bunker Hill Flour Mill. It is the last remaining mill in West Virginia with two waterwheels.

Five towering stories of limestone rock and faded red clapboard siding are located in a bend of the road from Bunker Hill. There were once about twelve mills on Mill Creek, between Gerrardstown and Middleway on the Opequon Creek. Only one other still stands and is now a residence.

The Bunker Hill Flour Mill was the last working mill on the stream. All the machinery is intact and operable. The Dehaven family owned it from 1936 until the late 1960s. Local residents remember, "Mr. Mac's buckwheat flour was the best around."

The Giles family moved to the large 1850s brick house across from the millpond in 1971. Paul Giles ran a retail feed business from September 1986 until July of 1995 (just before his death). His son, Steve, and the family continue to maintain the museum type atmosphere of the area. The Giles family collection of old farm machinery, steam engines, and related objects are housed in several buildings on the property.

Also on the property is a replica of a 1920s one room school house, built by Mr. Giles in the early 1990s. His fondest memories were of the one room school he attended east of Winchester, Virginia in the early 1940s. For more information call (304) 229-8707.

Story and photo by Janita Giles.

EASTON ROLLER MILL

WEST VIRGINIA

Easton Roller Mill was built in 1864 in the community of Easton, east of Morgantown, West Virginia. The mill originally served as a steam powered gristmill and sawmill. Two or more sets of millstones were used to produce custom ground grains for the surrounding community the first thirty years of operation. During peak operation, the mill ran day and night, grinding alternately flour and animal feed. In 1883, 120 bushels of grain were ground per day.

In 1894, roller milling technology replaced all but one set of stone mills. The "new" milling process required flour dressers, bran dusters, sieve scalpers, and several other pieces of associated equipment. These were added to the corn shellers, corn and cob crushers, and sieves already there. The mill continued operation, making flour and various feed mixtures until it closed permanently in 1940. Removal of the sawmill took place around 1920.

The mill structure is two and one-half stories high of post and beam construction with hand hewn chestnut and oak timbers connected by pinned mortise and tendon joints. The facade is the gothic style popular in that day with sawn cornice brackets on alternating centers and vertical board and batten siding. It stands today in a remarkable state of preservation as an example of both ancient and "modern" (circa 1900) milling practices.

The Easton Roller Mill, owned by the Monongalia Historical Society, is not operational although most of the original milling machinery is present. For more information call (304) 599-4393.

Story and photo courtesy of John Hamilton.

Glade Creek Gristmill
West Virginia

Autumn at Babcock's Glade Creek Gristmill is a colorful display of nature's dazzling fireworks. The Glade Creek Gristmill is a "new" mill completed in 1976 at Babcock. Fully operable, the mill is a re-creation of one that once ground grain on Glade Creek long before Babcock became a state park. Known as Cooper's Mill, the original mill stood on the present location of the park's Administration Building parking lot.

The mill consists of parts and pieces from several old mills which once dotted the state. The basic structure of the mill came from the Stoney Creek Gristmill, dating back to 1890, and was dismantled and moved piece by piece from a spot near Campbelltown in Pocahontas County. After an accidental fire destroyed the Spring Run Gristmill near Petersburg, Grant County, only the overshot water wheel could be salvaged. It now is a part of the Glade Creek Mill. Several other parts came from the Roaring Creek Gristmill near Seneca Rocks and Onego, Pendleton County.

A living monument to the over five hundred mills which thrived in West Virginia at the turn of the century, the Glade Creek Gristmill grinds cornmeal and buckwheat flour. The mill takes you back to the time when grinding grain by a rushing stream was a way of life and the groaning mill wheel was music to the miller's ear. For more information call 1-800-CALL WVA.

Courtesy of Babcock State Park.
Photo by J.C. Barger.

LYONS MILL
WEST VIRGINIA

John Day built Lyons Mill in Williamsport, West Virginia in 1882.

In 1894 John McDonald purchased the mill and is credited with installing the Midget Marvel, a type of steel roller mill. The stone buhrs were still used for grinding corn and buckwheat.

In 1904 the mill sold to Acey Ault who ran it until 1907, then sold to David Cassidy. Cassidy operated it twenty years, until his death.

David Lyon, the next owner, passed it on to his son, Arnold. Although the mill has changed hands three times since then, it is still known as Lyons Mill. The mill is currently owned by George Padgett.

Lyons Mill, the only one in Grant County, is higher in the mountains than most. The Fitz steel wheel measuring twenty-four feet in diameter and three feet wide, is the second largest in West Virginia, and powered by a spring fed stream 1,400 feet away.

At one time, the mill also served as a power plant suppling power to Williamsport and surrounding houses. Upgrades were made to the mill as needed. The original wooden wheel was replaced in the 1930s. The original stone buhrs were replaced with a more sanitary Meadows Mill. During WWII, the mill ran twenty-four hours a day and supplied over a barrel an hour for the war effort.

According to documents from the Rockingham (Va.) Recorder, the diary of W.H. Arehart, Sixty-Seventh Virginia Calvary, tells of the burning of the mill in Williamsport by Union troops, on Monday Jan. 4, 1864. Rather than leave the mill and store of grain for the Confederates, they torched it and the surrounding buildings. It is said that this mill is built on the original foundation.

As in most towns, the mill served as the gathering place for the men of the area, who would often travel here to have their grain ground and would take time to trout fish while they waited. Plans are to recreate this same setting by opening the mill and other buildings as a living museum.

Story and photo courtesy George R. Padgett and Carolyn Wright.

McCoy's Mill
West Virginia

McCoy's Mill is a secluded retreat, embracing the tranquility offered by the majestic mountains of Pendleton County. One of the oldest landmarks in West Virginia, where the Thorn Creek rushes into the South Branch of the Potomac, the first mill was built on this site in 1757.

The Indians called the South Branch Wappatomika, which meant River of Wild Geese, according to Elsie Byrd Boggs' book, *"A History of Franklin - The County Seat of Pendleton Co., West Virginia"*. Since the mill's beginning the mouth of the Thorn has been the center of activity.

The first white man to come here was John Vanmeter, a Dutch trader from New York State. He arrived with a band of Delaware Indians in 1735 in a raid against the Catawbas. They were defeated at the mouth of the Thorn and retreated down the valley.

By 1747 immigration had begun in Pendleton County. Ulrich Conrad, Sr., who came from Switzerland in 1753, built a gristmill at the mouth of the Thorn in 1757. He furnished military supplies for the soldiers of Dunmore's War in 1774 and the Revolutionary War in 1782. Gunpowder was made in the area and saltpeter mined in a nearby cave during the Revolutionary and Civil Wars.

The mill descended to Ulrich Conrad, Jr. and sold to General William McCoy, a prosperous farmer, a member of Congress and friend of General Stonewall Jackson. The mill passed to his nephew, William McCoy, who served in the State Legislature. Milling became so profitable that he replaced the old mill for one of greater capacity in about 1845. He found an expert millwright, James Wm. Byrd, from the Shenandoah Valley of Virginia. He finished the basement floor and walls with plaster, made by hand of lime, sand and hair. He carved his initials, JWB, in the wall and they are still there today. This history is taken from *"A History of Franklin - The County Seat of Pendleton Co., West Virginia"*, by Elsie Boggs.

Who knows how many floods have attempted to destroy this historic building and haven't succeeded? The flood of 1985 damaged the mill, but repairs were made.

Glen Hofecker found and purchased the mill while working with the Church of the Brethren disaster program. They had come after the 1985 flood to build swinging bridges. Having forgotten the hard work it had taken to restore the Forestville (Zirkle) Mill in Virginia, he began restoration. He built a log house, incorporating the chimney of the old two-story log house used by the miller. This is used as his workshop, where he builds Chippendale furniture. The mill has changed owners many times and had numerous uses since the 1900s, when operations ceased. The Miller's house is now a Bed and Breakfast. For more information call (304) 358-7893.

Story and photo courtesy Glen and Iris Hofecker, owners.

Beckman Mill
Wisconsin

In 1844, Charles F.H. Goodhue, Junior, a prominent businessman in Beloit, Wisconsin, purchased property in rural Rock County on the west branch of Raccoon Creek. There he built a whiskey distillery. It burned around 1852. In 1868 William How, a miller from Beaver Dam, Wisconsin, purchased the site at sheriff's auction and built the mill that stands today. In 1874 the mill sold to William T. Kelley.

Catherine Beckmann bought the property in 1882, and her husband, German-born August, became the miller. August operated the mill until his death in 1908, when his sons Henry and Charlie Beckman bought out Herman Griinke, August's son-in-law and partner since 1894. The early 1900s saw the mill's peak of production. Wagons filled with grain lined up in the road, and the Beckman brothers mill ground day and night. Well known locally for its buckwheat flour, business was booming. The dam was rebuilt and the building enlarged to include a drive-through and a shed to house an auxiliary engine that supplemented the turbine power.

For many years, the mill served as an important gathering place for the surrounding rural community. It housed the Newark town post office from 1890 to 1892. Farmers gathered to have their grain ground while catching up on the news. A swimming hole, dance hall, creamery, and general store provided for social activities. Charlie Beckman, a fiddler and other family members regularly played for local dances.

However, business slowed during the Depression. Most of the area agriculture switched from wheat to corn. Improved roads opened broader markets for farm products and farmers no longer relied as much on the local miller. Henry died in 1939 and Charlie ceased milling operations in 1954. Charlie and Bessie Beckman raised five sons in their house near the mill, a building dating back to the time of the distillery. He died in 1966. Bessie continued to live in the house until 1978, when Rock County purchased the property for use as a park, intending to restore the mill, but the building continued to deteriorate. In 1990, alarmed at the mill's decay, a group of concerned area residents, historians, and Beckman family members formed the Friends of Beckman Mill. The new group leased two acres from the county and began the restoration.

On the verge of collapse, the mill was jacked up so new concrete footings could be poured and a new stone foundation laid. Restoration included roof repairs, flooring replacement, completion of the turbine pit, and reconstruction of the decayed additions, bringing the mill back to its 1920s appearance. In addition, two authentic power sources, the mill's original Leffel turbine and an International Harvester Titan tractor engine were rebuilt and brought back to running condition. Now, thanks almost entirely to the work of dedicated volunteers, the Beckman Mill operates as a working museum to exhibit a way of life and work that is nearly forgotten. For more information contact the Beckman Mill at (608) 368-1435.

Story by Steve Frevert. Photo by Kevin Johnson.

DELLS MILL

WISCONSIN

In the 1830s, early pioneers found Wisconsin's rich soil ideal for growing wheat, and hundreds of rivers and streams offered an abundant supply of water power as settlers poured into the area.

It was important to each farmer that there be a mill near his farm to grind the wheat into usable flour. Much activity centered around the mill. Farmers came to have their grain ground, travelers paused for food and lodging and immigrants stopped to ask for information. A visit to the mill became a pleasant occasion including a picnic lunch and fishing by the millpond.

The growth of wheat production in Wisconsin was tremendous between 1850 and 1880. In 1840, the area supported a total of thirty-three flour and gristmills and by 1880 just under 1,000. By 1900, milling had become the second largest industry in the state. Shortly after the turn of the century, the milling industry started its long decline and by 1920 abandoned flour mills could be found throughout the state. The record harvest of wheat over many years had depleted the soil and could no longer support big crops.

By the time of the Civil War the establishment of a community on the Dells Mill site was well under way.

An 1869 deed transferred the property from William Weber to Stephen Marston and lists the property as consisting of sixteen acres containing "the mill, boarding house, school house and other buildings thereon". Of these, only the mill remains.

Dells Mill operated as a flour and feed mill for over one hundred years, with Bridge Creek waters as the only source of power use.

In 1894 the mill was purchased by the Clark family. After four generations they still retain ownership. In 1968, Dells Mill became a museum, grinding on special occasions.

The mill building is a five story structure built of hand-hewn timbers and held together by wooden pegs. It contains 3,000 feet of belting and 175 pulleys, all powered by water. For more information call (715) 286-2714.

Photos and story courtesy of Gustave Clark.

Figor's Mill
Wisconsin

Just east of the village of Wyocena, Wisconsin, on the south branch of Duck Creek, stands Figor's Mill. It was built 1853-54 on the bank of the millpond by Darius Bisbee. The gristmill, struck by lightning, burned to the ground in 1910.

J.D. Figor rebuilt the mill with two Leffel turbines. It provided electrical power to neighbors and later to the local utility company until shut down in 1973.

Figor's Mill, refurbished and rebuilt by skilled family members and friends, is once again operational. It provides electrical energy, recreation and pleasure for many.

Story and photo courtesy Blayde and Linda Elert.

Krug Village Gristmill

Wisconsin

At the Fond du Lac County Historical Society's site, Galloway House and Village, is the Krug Gristmill. Its beauty is enhanced by the twelve foot overshot waterwheel.

Most mills were built in or by ravines, which had flowing water to fill the buckets. The weight of the water in the buckets turned the wheel. In the setup of the Krug Gristmill, the water in the pond is recycled. The water is pumped back up into the flume by means of a small electric pump. It then flows from the flume into the buckets.

The gears, shaft and grinding stones were from the Krug Mill of the Township of Empire, Fond du Lac County. The Treleven brothers, early community settlers, built the mill shortly after the Civil War. The wooden gears can be quickly replaced when broken or worn, an advantage over modern machinery.

The mill was reconstructed in 1966. Much of the work was accomplished through the efforts of H.A. Ladewig, who copied and built the overshot waterwheel from old designs and pictures. He also provided the mechanical skill to link the power from the waterwheel to the milling stones and chaff fan. He continued to maintain the mill for many years.

On the west wall of the second floor is a diagram of the grinding stones. On the east wall is a map of the Fond du Lac River and the many mills along the river during the early days of the City.

This is one of the working exhibits in the village. Most of the corn that is ground is returned for farm feed. For more information, call (920) 922-6390.

Story and photo courtesy of Rosalie Luedtke, Fond Du Lac County Historical Society. Photo by Ray Thornton.

The Old Feed Mill

Wisconsin

The Old Feed Mill's rich history dates back to 1857, the year Mazomanie became platted as a village. The water-powered Lynch and Walker Flouring Mill grew to be a large regional mill. By the turn of the century it was the leading driver of commerce in a thriving Mazomanie.

During the middle half of the twentieth century, highways grew in popularity and railroads serving the mill were not as important. Other mills began to dominate and the Old Feed Mill shifted to grinding cattle feed for local farmers. The community lost its robustness as people went elsewhere for employment.

In the late 1980s the community began to reassert itself, as did The Old Feed Mill. Then in the winter of 1992, Dan and Nancy Viste gazed upon the abandoned, sun-warmed stone mill and saw exciting potential. The mill had just been placed on the National Register of Historic Places along with thirty-three other downtown historic structures. The Vistes turned the crumbling mill into an award winning restoration, receiving a prestigious Certificate of Commendation from the Wisconsin State Historical Society. The mill now features a restaurant, gift shop, bakery and has teamed with Wisconsin Southern Railroad to provide rail service form Madison to Mazomanie.

Motivated by their respect for heritage and basic Midwestern values the Vistes have restored the building and the sense of tradition its history represents. For more information call (888) 345-4909.

Story and photo courtesy of Dan and Nancy Viste. Photo by Marilyn Collins.

Nemahbim Roller Mills

Wisconsin

This site dates from a time when milling and power generation faced a great juncture between old ideas and new ideas, between small mills serving a local community and mills of vast power supplying grain products to the world.

When the 1836 government survey of Wisconsin Territory public domain lands reached completion, the human floodtide began. Armed with survey maps, the settlers and speculators came looking for agricultural land, timberlands, and hydropower sites.

One such settler, Dr. Milton Buckingham Cushing, had family roots back to the Mayflower. Suffering financial distress in the Panic of 1837 he eagerly took advantage of the new territory. In the mid-1830s he laid claim to 320 acres of oak savannah. Hidden in the southeastern lake country ran the little Bark River. Below a forty-eight square mile watershed he spotted the ideal mill site, a rapidly moving river one mile downstream from a large lake. The Bark River, flowing through a large lake would have even flow, with water available even in dry seasons and flash floods contained on the large surface. Better yet, the territorial legislature surveyed and had the financial resources to construct a canal on his stretch of the river.

Dr. Cushing became consumptive and had to abandon his dream. He sold sixty acres of this great power potential to another visionary, Albert Alden, also a Plymouth Rock escapee to the frontier. Alden contracted with Nelson Page Hawks the construction of a mill and dam until the Milwaukee-Rock River Canal project went through to supply waterpower.

Nelson Page Hawks, a great entrepreneur, had business interests in a wide area. A Greek Revival-style hotel he built twenty-five miles west of Milwaukee and close to the site of the proposed Milwaukee-Rock River Canal served as his headquarters. According to contract with Albert Alden, he built Telegraph Mills on the north bank of the river. His mill, unfortunately, burned Dec. 23, 1853. The fire was of suspicious origin.

Despite a complete loss, Hawks rebuilt immediately, moving the site of his new mill, Twin Mills, to the south bank of the river, as the Milwaukee-Rock River Canal project had been abandoned. Twin Mills is the building seen today in Delafield.

Built of local oak, tamarack and maple, Twin Mills has twelve inch hand-hewn beams and joists and rafters of tamarack. Flooring consisted of random width eleven and one half inch thick soft woods. The four-story building (counting the walk-in basement) featured a one-man elevator operating on counterbalance. With three runs of stone he did a roaring business, as reported in the seventh federal census. Records of other equipment have not been found. However, Hawks had been dickering with an Ohio firm for a Zebulon & A. Parker Percussion & Reaction Waterwheel. It is not known the fate of the wheel, whether installed and destroyed in the first mill or operated in Twin Mills. Hawks died ten years later in 1863, and the mill sold to William Notbohm, who added a two-story storage wing south of the main building. When he died, his widow sold to her employee, George Apfelbacher, who learned the trade in Germany. In the 1870s the name of the mill changed to Nemahbin Roller Mills. Milling became unprofitable for Apfelbacher and during the World War II scrap drives, he scraped the irons and operated a rooming house.

Today the Nemahbin Roller Mills perch like a sentinel on its 400 foot dam. It sits serenely in the sun like an ancient dowager, contemplating her picturesque pond. Hidden in its depths is the little Bark River, measured at thirty links width in 1836. The settlement around her pond has completely obliterated the ancient oak savannah. Northeast at the beginning of her impoundment is the Cushing Memorial erected eighty-five years ago by the people of the nation to the heroic sons of that first family which had such dreams of power generation. To the east is the museum honoring the Hawks family. The steady hum of traffic from the great interstate highway, which took the place of the Milwaukee-Rock River Canal, can be heard. From her depths comes a reassuring purr of co-generation.

In the words of a famed Wisconsin power engineer: "As to heritage, we can leave dollars, or we can leave a hydroelectric plant which will continue generation after generation."

Story courtesy Margaret Zerwekh.

Nemahbim Roller Mills
Photo courtesy Margaret Zerwekh.

Rock Mill
Wisconsin

The building of Rock Mill took place in 1847, the year before Wisconsin attained statehood. Rock Mill sits in a scenic valley carved into the Niagara Escarpment, with its magnificent limestone outcroppings and hilly terrain. Through this valley and over the layers of limestone flow the cascading waters of the Devils River, which powered the mill. Rock Mill, the oldest existing historic mill in Manitowoc County, has most of the original milling equipment intact. The county once boasted thirty-eight waterpowered mills.

Originally built as a sawmill by English millwright and speculator Pliney Pierce, Rock Mill soon passed through a series of owners. One of the owners converted it to a gristmill where wheat (or graham), buckwheat and rye flours were ground.

In 1878, Emil Schlegelmilch, a German immigrant and fifth generation miller, became the proud owner of Rock Mill and all its equipment. With the mill, he also acquired a barn, some cattle, a home (complete with furnishings) and the eighty acres on which these structures stood for $3500.

Schlegelmilch's Rock Mill operated using two runs of French buhrstones, a hand-crafted wooden auger and many other hand-hewn pulleys, lathes and tools. The mill also served as the local social center, post office, hardware store and precinct voting place.

Rock Mill is along what was once the Military Road from Fort Dearborn in Chicago to Fort Howard in Green Bay. Military personnel frequently used the mill site as an encampment. The falls and rapids of the Devils River flow gently for most of the year but in the spring can become quite turbulent. Legend has it that early in 1827, the rushing waters swept young Lt. Zachary Taylor off his horse and into the river. Fortunately, his men rescued him and he went on to become the twelfth President of the United States.

Emil Schlegelmilch died in March of 1934 and within days the dam on Devils River was washed out by huge ice floes set loose by the spring thaws. Rock Mill ceased to operate and gradually fell into a state of serious disrepair.

Some thirty years later, the Robert Lyman family bought Rock Mill and began immediate restoration of the mill and adjacent buildings. Lyman dedicated intense personal effort to the project. In June of 1982 Rock Mill was placed on the National Registry of Historic Places for both its architectural and historical significance. Rock Mill is now a museum. For more information call (920) 863-2812.

Story and photo courtesy Robert & Doris Lyman.

Alphabetical Index

H

I

J

K

L

M

N

O

P

R

S

T

U

V

W

Y

Z

State Index

Minnesota

Mississippi

Missouri

Nebraska

New Hampshire

New Jersey

New Mexico

New York

North Carolina

Ohio

Oregon

Pennsylvania

Rhode Island

South Carolina

South Dakota

Tennessee

Texas

Utah

Virginia

Washington

West Virginia

Wisconsin